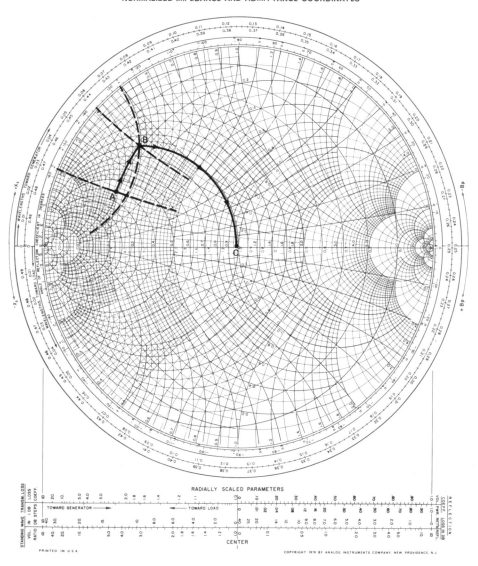

Figure 2.4.8b

# MICROWAVE TRANSISTOR AMPLIFIERS
# Analysis and Design

### Guillermo Gonzalez, Ph.D.
*Professor of Electrical and Computer Engineering*
*University of Miami*

PRENTICE-HALL, INC., Englewood Cliffs, N.J. 07632

*Library of Congress Cataloging in Publication Data*

GONZALEZ, GUILLERMO, 1944–
  Microwave transistor amplifiers.

  Bibliography: p.
  Includes index.
  1. Transistor amplifiers.  2. Microwave amplifiers.
I. Title.
TK7871.2.G59  1984      621.381′33      84-2074
ISBN  0-13-581646-7

Editorial/production supervision
  and interior design: *Theresa A. Soler*
Manufacturing buyer: *Anthony Caruso*

© 1984 by Prentice-Hall, Inc., Englewood Cliffs, New Jersey 07632

All rights reserved. No part of this book may be
reproduced, in any form or by any means,
without permission in writing from the publisher.

Printed in the United States of America

10  9  8  7  6  5  4  3

ISBN 0-13-581646-7

Prentice-Hall International, Inc., *London*
Prentice-Hall of Australia Pty. Limited, *Sydney*
Editora Prentice-Hall do Brasil, Ltda., *Rio de Janeiro*
Prentice-Hall Canada Inc., *Toronto*
Prentice-Hall of India Private Limited, *New Delhi*
Prentice-Hall of Japan, Inc., *Tokyo*
Prentice-Hall of Southeast Asia Pte. Ltd., *Singapore*
Whitehall Books Limited, *Wellington, New Zealand*

# CONTENTS

**PREFACE** *vii*

## 1 — REPRESENTATIONS OF TWO-PORT NETWORKS  1

| | | |
|---|---|---:|
| 1.1 | Introduction | *1* |
| 1.2 | The Impedance, Admittance, Hybrid, and *ABCD* Matrices | *1* |
| 1.3 | Traveling Waves and Transmission-Line Concepts | *4* |
| 1.4 | The Scattering Matrix and the Chain Scattering Matrix | *8* |
| 1.5 | Shifting Reference Planes | *12* |
| 1.6 | Properties of Scattering Parameters | *13* |
| 1.7 | Generalized Scattering Parameters | *19* |
| 1.8 | Two-Port Network Parameters Conversions | *22* |
| 1.9 | Scattering Parameters of Transistors | *23* |
| 1.10 | Characteristics of Microwave Transistors | *31* |

## 2 — MATCHING NETWORKS AND SIGNAL FLOW GRAPHS  42

| | | |
|---|---|---:|
| 2.1 | Introduction | *42* |
| 2.2 | The Smith Chart | *43* |
| 2.3 | The Normalized Impedance and Admittance Smith Chart | *51* |

| | | |
|---|---|---:|
| 2.4 | Impedance Matching Networks | *55* |
| 2.5 | Microstrip Matching Networks | *67* |
| 2.6 | Signal Flow Graph | *80* |
| 2.7 | Applications of Signal Flow Graphs | *84* |

## 3 — MICROWAVE TRANSISTOR AMPLIFIER DESIGN     *91*

| | | |
|---|---|---:|
| 3.1 | Introduction | *91* |
| 3.2 | Power Gain Equations | *92* |
| 3.3 | Stability Considerations | *95* |
| 3.4 | Constant-Gain Circles—Unilateral Case | *102* |
| 3.5 | Unilateral Figure of Merit | *111* |
| 3.6 | Simultaneous Conjugate Match—Bilateral Case | *112* |
| 3.7 | Constant-Gain Circles—Bilateral Case | *114* |
| 3.8 | Operating and Available Power Gain Circles | *119* |
| 3.9 | DC Bias Networks | *125* |

## 4 — NOISE, BROADBAND, AND HIGH-POWER DESIGN METHODS     *139*

| | | |
|---|---|---:|
| 4.1 | Introduction | *139* |
| 4.2 | Noise in Two-Port Networks | *140* |
| 4.3 | Constant Noise Figure Circles | *142* |
| 4.4 | Broadband Amplifier Design | *154* |
| 4.5 | Amplifier Tuning | *169* |
| 4.6 | Bandwidth Analysis | *170* |
| 4.7 | High-Power Amplifier Design | *174* |
| 4.8 | Two-Stage Amplifier Design | *187* |

## 5 — MICROWAVE TRANSISTOR OSCILLATOR DESIGN     *194*

| | | |
|---|---|---:|
| 5.1 | Introduction | *194* |
| 5.2 | One-Port Negative-Resistance Oscillators | *194* |
| 5.3 | Two-Port Negative-Resistance Oscillators | *199* |
| 5.4 | Oscillator Design Using Large-Signal Measurements | *203* |
| 5.5 | Oscillator Configurations | *208* |

**Appendix**

**— A —** COMPUTER AIDED DESIGN:
COMPACT AND SUPER-COMPACT       *217*

**Appendix**

**— B —** UM-MAAD       *230*

INDEX       *241*

# PREFACE

This book presents a unified treatment of the analysis and design of microwave transistor amplifiers using scattering parameters techniques. The term *microwave frequencies* is used to refer to those frequencies whose wavelengths are in the centimeter range (i.e., 1 to 100 cm). However, the design procedures and analyses presented in the book are not limited to the microwave frequencies. In fact, they can be used in any frequency range where the scattering parameters of a transistor are given.

The transistors used in microwave amplifiers are the bipolar junction transistor (BJT) and the gallium arsenide field-effect transistor (GaAs FET). The BJT performs very well up to approximately 4 GHz. In this frequency range the BJTs are reliable, low cost, (usually) unconditionally stable, have a high gain, and a low noise figure.

The GaAs FET performance above, approximately 4 GHz, is superior to that of the BJT. For example, with a GaAs FET a noise figure of 1.5 dB with a 20-dB gain can be obtained at 4 GHz, and a noise figure of 2.7 dB with 10-d B gain can be obtained at 18 GHz. Power GaAs FETs are capable of several-watt operation above 4 GHz. At the present time, GaAs FET amplifiers are ready to take the next step in frequency from 20 to 40 GHz [1].

Microwave transistors are represented by two-port networks and are characterized by scattering parameters. The scattering parameters are popular because they are easy to measure with modern network analyzers, their use in microwave transistor amplifier design is conceptually simple, and they provide meaningful design information. Furthermore, flow graph theory is readily applicable.

Chapters 1 to 5 present the basic principles and techniques used in microwave transistor amplifier analysis and design. These chapters provide the foundation for a well-designed microwave transistor amplifier. An introduction to computer-aided design (CAD) using large-scale programs is given in Appendix A. It is only after the problem is fully understood that CAD techniques should be used. Otherwise, the natural human tendency to be erroneous and inefficient can substantially increase the cost of a design. The large-scale CAD programs used in Appendix A are COMPACT and SUPER-COMPACT, trademarks of Compact Software, Inc. [2].

In Appendix B a listing of the program UM-MAAD (University of Miami-Microwave Amplifiers Analysis and Design) is given. This CAD program, written in FORTRAN-77, is simple to use and understand. It can be used in the analysis and design of some microwave transistor amplifiers. The reader will find this program very useful in the solution of the examples and problems in the book, especially those from Chapters 3 to 5.

Another inexpensive set of CAD programs is available in the text Microwave Circuit Design Using Programmable Calculators by Allen and Medley [3]. Also, CAD programs suitable for mini-computers and programmable calculators are available from Compact Software, Inc. [2].

This book was based on the notes developed for a senior-graduate level course in microwave transistor amplifiers at the University of Miami. The original set of notes was based mainly on the Hewlett-Packard Application Note 154 [4] and the text by Carson [5]. Subsequently, these notes were greatly expanded using the vast source of information that appeared in technical and professional journals. The reference sections list the principal references used.

This book is intended to be used in a senior-graduate level course in microwave transistor amplifiers or by practicing microwave engineers. It is assumed that the reader has completed the undergraduate network theory, electronics, and electromagnetic courses, or equivalent courses. The transmission-line theory needed is fully covered in the book, especially the use of the Smith chart as a design tool.

The problems at the end of each chapter form an integral part of the text, and even if they are not solved, they should be read.

I wish to thank Mr. Les Besser for introducing me to COMPACT and for providing the Department of Electrical Engineering at the University of Miami with a copy of COMPACT for teaching and research purposes. Furthermore, I thank Mr. Besser for working the example in Appendix A using SUPER-COMPACT.

I also wish to thank all of my former students for their helpful comments, especially, the invaluable suggestions and constructive criticisms from Avanic Branko, William Sanfiel, Deniz Ergener, Ching Y. Kung, Claudio J. Traslavina, and Levent Y. Erbora. The contributions of Avanic Branko to the solutions of the chapter problems were very helpful. The support that I have always received from Dr. Kamal Yacoub, Chairman of the Department of

Electrical and Computer Engineering at the University of Miami is greatly appreciated. To my colleagues, Professors Manuel A. Huerta, James C. Nearing, and Tsay Young my sincere gratitude for their help in many areas of science and most of all for their friendship. Finally, my deepest appreciation to my wife Pat, my children Donna and Alex, and my parents Ricardo and Raquel for their love, encouragement, and patience.

The peaceful surroundings of Fleetwood Falls, Ashe County, North Carolina, provided the needed serenity to complete this book.

*Guillermo Gonzalez, Ph.D*

## REFERENCES

[1] J. C. Rosenberg, G. J. Policky, and N. K. Osbrinke, "GaAs FET Amplifiers Reach Millimeter Waves," Microwaves, June 1982.

[2] COMPACT and SUPER-COMPACT (computer programs), Compact Software, Inc., 1131 San Antonio Road, Palo Alto, California 94303.

[3] J. L. Allen and M. W. Medley Jr., *Microwave Circuit Design Using Programmable Calculators*, Artech House, Inc., Dedham, Massachusetts, 1981.

[4] "S Parameter Design," Hewlett-Packard Application Note 154, April 1972.

[5] R. S. Carson, *High-Frequency Amplifiers*, Wiley-Interscience, New York, 1975.

# 1
# REPRESENTATIONS OF TWO-PORT NETWORKS

## 1.1 INTRODUCTION

In order to characterize the behavior of a two-port network, measured data of both its transfer and impedance functions must be obtained. At low frequencies, the $z$, $y$, $h$, or $ABCD$ parameters are examples of network functions used in the description of two-port networks. These parameters cannot be measured accurately at higher frequencies because the required short- and open-circuit tests are difficult to achieve over a broadband range of microwave frequencies.

A set of parameters that is very useful in the microwave range are the *scattering parameters* ($S$ parameters). These parameters are defined in terms of traveling waves and completely characterize the behavior of two-port networks at microwave frequencies.

In the 1970s the popularity of $S$ parameters increased because of the appearance of new network analyzers, which performed $S$-parameter measurements with ease. The $S$ parameters are simple to use in analysis, and flow graph theory is directly applicable. Although the principal use of $S$ parameters in this text is in the characterization of two-port networks, they can also be used in the characterization of $n$-port networks.

## 1.2 THE IMPEDANCE, ADMITTANCE, HYBRID, AND *ABCD* MATRICES

At low frequencies the two-port network shown in Fig. 1.2.1 can be represented in several ways. The most common representations are the impedance

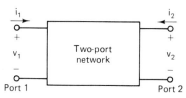

**Figure 1.2.1** Two-port network representation.

matrix ($z$ parameters), the admittance matrix ($y$ parameters), the hybrid matrix ($h$ parameters), and the chain or *ABCD* matrix (chain or *ABCD* parameters). These parameters are defined as follows:

*z Parameters:*

$$v_1 = z_{11}i_1 + z_{12}i_2$$
$$v_2 = z_{21}i_1 + z_{22}i_2$$

or in matrix form

$$\begin{bmatrix} v_1 \\ v_2 \end{bmatrix} = \begin{bmatrix} z_{11} & z_{12} \\ z_{21} & z_{22} \end{bmatrix} \begin{bmatrix} i_1 \\ i_2 \end{bmatrix}$$

*y Parameters:*

$$\begin{bmatrix} i_1 \\ i_2 \end{bmatrix} = \begin{bmatrix} y_{11} & y_{12} \\ y_{21} & y_{22} \end{bmatrix} \begin{bmatrix} v_1 \\ v_2 \end{bmatrix}$$

*h Parameters:*

$$\begin{bmatrix} v_1 \\ i_2 \end{bmatrix} = \begin{bmatrix} h_{11} & h_{12} \\ h_{21} & h_{22} \end{bmatrix} \begin{bmatrix} i_1 \\ v_2 \end{bmatrix}$$

*ABCD Parameters:*

$$\begin{bmatrix} v_1 \\ i_1 \end{bmatrix} = \begin{bmatrix} A & B \\ C & D \end{bmatrix} \begin{bmatrix} v_2 \\ -i_2 \end{bmatrix}$$

The previous two-port representations are very useful at low frequencies because the parameters are readily measured using short- and open-circuit tests at the terminals of the two-port network. For example,

$$z_{11} = \left. \frac{v_1}{i_1} \right|_{i_2=0}$$

is measured with an ac open circuit at port 2 (i.e., $i_2 = 0$).

The $z$, $y$, and *ABCD* parameters are also useful in the computer analysis of circuits. When two-port networks are connected in series, as shown in Fig. 1.2.2, we can find the overall $z$ parameters by adding the individual $z$ parame-

## Sec. 1.2 The Impedance, Admittance, Hybrid, and ABCD Matrices

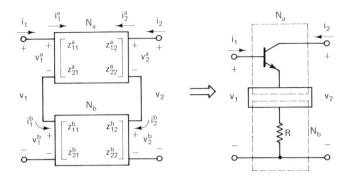

**Figure 1.2.2** Series connection using $z$ parameters and a typical application.

ters, namely

$$\begin{bmatrix} v_1 \\ v_2 \end{bmatrix} = \begin{bmatrix} v_1^a + v_1^b \\ v_2^a + v_2^b \end{bmatrix} = \begin{bmatrix} z_{11}^a + z_{11}^b & z_{12}^a + z_{12}^b \\ z_{21}^a + z_{21}^b & z_{22}^a + z_{22}^b \end{bmatrix} \begin{bmatrix} i_1 \\ i_2 \end{bmatrix}$$

When two-port networks are connected in shunt, as shown in Fig. 1.2.3, we can find the overall $y$ parameters by adding the individual $y$ parameters, namely

$$\begin{bmatrix} i_1 \\ i_2 \end{bmatrix} = \begin{bmatrix} i_1^a + i_1^b \\ i_2^a + i_2^b \end{bmatrix} = \begin{bmatrix} y_{11}^a + y_{11}^b & y_{12}^a + y_{12}^b \\ y_{21}^a + y_{21}^b & y_{22}^a + y_{22}^b \end{bmatrix} \begin{bmatrix} v_1 \\ v_2 \end{bmatrix}$$

When cascading two-port networks the chain or $ABCD$ matrix can be used as follows (see Fig. 1.2.4):

$$\begin{bmatrix} v_1 \\ i_1 \end{bmatrix} = \begin{bmatrix} v_1^a \\ i_1^a \end{bmatrix} = \begin{bmatrix} A^a & B^a \\ C^a & D^a \end{bmatrix} \begin{bmatrix} v_2^a \\ -i_2^a \end{bmatrix} = \begin{bmatrix} A^a & B^a \\ C^a & D^a \end{bmatrix} \begin{bmatrix} A^b & B^b \\ C^b & D^b \end{bmatrix} \begin{bmatrix} v_2^b \\ -i_2^b \end{bmatrix} \quad (1.2.1)$$

because $v_2^a = v_1^b$ and $-i_2^a = i_1^b$. The relation (1.2.1) shows that the overall $ABCD$ matrix is equal to the product (i.e., matrix multiplication) of the individual $ABCD$ matrices.

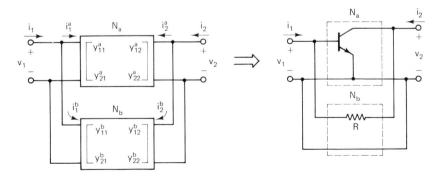

**Figure 1.2.3** Shunt connection using $y$ parameters and a typical application.

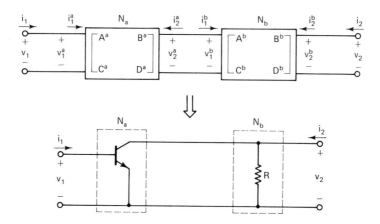

**Figure 1.2.4** Cascade connection using *ABCD* parameters and a typical application.

At microwave frequencies the *z*, *y*, *h*, or *ABCD* parameters are very difficult (if not impossible) to measure. The reason is that short and open circuits to ac signals are difficult to implement over a broadband, at microwave frequencies. Also, an active two-port (e.g., a microwave transistor) might oscillate under short- or open-circuit conditions. Therefore, a new representation of the two-port network at microwave frequencies is needed. The appropriate representation is called the *scattering matrix* and the scattering parameters are defined in terms of traveling waves.

## 1.3 TRAVELING WAVES AND TRANSMISSION-LINE CONCEPTS

The voltage and current along a transmission line are functions of position and time. For sinusoidal excitation, the instantaneous voltage and current can be expressed in the form

$$v(x, t) = \text{Re}\,[V(x)e^{j\omega t}]$$

and

$$i(x, t) = \text{Re}\,[I(x)e^{j\omega t}]$$

where Re means the real part. The complex quantities $V(x)$ and $I(x)$ are phasors and express the variations of the voltage and current as a function of position along the transmission line. The polar forms of a complex number $re^{j\theta}$ and $r\underline{/\theta}$ are both used in this book.

The differential equations satisfied by the phasors $V(x)$ and $I(x)$ along a

### Sec. 1.3 Traveling Waves and Transmission-Line Concepts

uniform transmission line are (see Problem 1.1)

$$\frac{d^2V(x)}{dx^2} - \gamma^2 V(x) = 0 \qquad (1.3.1)$$

and

$$\frac{d^2I(x)}{dx^2} - \gamma^2 I(x) = 0 \qquad (1.3.2)$$

where the complex propagation constant $\gamma$ is given by

$$\gamma = \alpha + j\beta = \sqrt{(R + j\omega L)(G + j\omega C)}$$

The attenuation constant $\alpha$ is given in nepers per meter and the propagation constant $\beta$ in radians per meter. The parameters $R$, $G$, $L$, and $C$ are the resistance, conductance, inductance, and capacitance per unit length of the transmission line. They are assumed to be constant along the transmission line (i.e., the transmission line is uniform).

The general solutions of (1.3.1) and (1.3.2) are

$$V(x) = Ae^{-\gamma x} + Be^{\gamma x} \qquad (1.3.3)$$

and

$$I(x) = \frac{A}{Z_o} e^{-\gamma x} - \frac{B}{Z_o} e^{\gamma x} \qquad (1.3.4)$$

where

$$Z_o = \sqrt{\frac{R + j\omega L}{G + j\omega C}}$$

is known as the *complex characteristic impedance* of the transmission line. The constants $A$ and $B$ are, in general, complex quantities.

Equations (1.3.3) and (1.3.4) represent the voltage and current along the transmission line as a pair of waves traveling in opposite directions, with phase velocity $v_p = \omega/\beta$ and decreasing in amplitude according to $e^{-\alpha x}$ or $e^{\alpha x}$. The wave $e^{-\gamma x} = e^{-\alpha x} e^{-j\beta x}$ is called the *incident wave* (outgoing wave) and the wave $e^{\gamma x} = e^{\alpha x} e^{j\beta x}$ is called the *reflected wave* (incoming wave). The quantity $\beta x$ is known as the *electrical length* of the line.

A transmission line of characteristic impedance $Z_o$ terminated in a load $Z_L$ is shown in Fig. 1.3.1. The reflection coefficient $\Gamma(x)$ is defined as

$$\Gamma(x) = \frac{Be^{\gamma x}}{Ae^{-\gamma x}} = \frac{B}{A} e^{2\gamma x} = \Gamma_o e^{2\gamma x} \qquad (1.3.5)$$

where $\Gamma_o$ is the load reflection coefficient, namely

$$\Gamma_o = \Gamma(0) = \frac{B}{A}$$

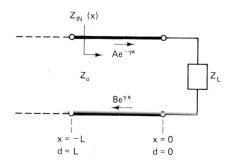

**Figure 1.3.1** Transmission line terminated in the load $Z_L$.

Therefore, the reflected wave can be written as $A\Gamma_0 e^{\gamma x}$ and (1.3.3) and (1.3.4) can be expressed in the form

$$V(x) = A(e^{-\gamma x} + \Gamma_0 e^{\gamma x})$$
$$I(x) = \frac{A}{Z_o}(e^{-\gamma x} - \Gamma_0 e^{\gamma x})$$
(1.3.6)

The input impedance of the transmission line at any position $x$ is defined as

$$Z_{IN}(x) = \frac{V(x)}{I(x)} = Z_o \frac{e^{-\gamma x} + \Gamma_0 e^{\gamma x}}{e^{-\gamma x} - \Gamma_0 e^{\gamma x}} \qquad (1.3.7)$$

where the constant $\Gamma_0$ can be evaluated using the condition

$$Z_{IN}(0) = Z_L$$

Then

$$Z_L = Z_o \frac{1 + \Gamma_0}{1 - \Gamma_0}$$

or

$$\Gamma_0 = \frac{Z_L - Z_o}{Z_L + Z_o} \qquad (1.3.8)$$

Equation (1.3.8) shows that $\Gamma_0 = 0$ when $Z_L = Z_o$. That is, there is no reflection in a properly terminated or matched (i.e., $Z_L = Z_o$) transmission line. Substituting (1.3.8) into (1.3.7) and letting $x = -d$ gives

$$Z_{IN}(d) = Z_o \frac{Z_L + Z_o \tanh \gamma d}{Z_o + Z_L \tanh \gamma d} \qquad (1.3.9)$$

The change $x = -d$ is normally done in transmission-line problems in order to measure positive distances as one moves from the load toward the source.

At microwave frequencies $R$ and $G$ are usually negligible and the trans-

## Sec. 1.3 Traveling Waves and Transmission-Line Concepts

mission line is said to be lossless. In a lossless transmission line

$$\alpha = 0$$
$$\gamma = j\beta$$
$$\beta = \omega\sqrt{LC}$$
$$v_p = \frac{1}{\sqrt{LC}}$$
$$\lambda = \frac{v_p}{f}$$
$$V(x) = Ae^{-j\beta x} + Be^{j\beta x}$$
$$I(x) = \frac{A}{Z_o} e^{-j\beta x} - \frac{B}{Z_o} e^{j\beta x}$$

and

$$Z_o = \sqrt{\frac{L}{C}}$$

Observe that $Z_o$ is real. Also, from (1.3.9) the input impedance in a lossless transmission line can be expressed in the form

$$Z_{IN}(d) = Z_o \frac{Z_L + jZ_o \tan \beta d}{Z_o + jZ_L \tan \beta d} \quad (1.3.10)$$

Unless otherwise specified, all transmission lines in this book are assumed to be lossless and uniform.

The two waves traveling in opposite directions in a transmission line produce a standing-wave pattern. From (1.3.6), the maximum value of the voltage along the line has the value

$$|V(x)|_{max} = |A|(1 + |\Gamma_o|)$$

and the minimum value of the voltage is

$$|V(x)|_{min} = |A|(1 - |\Gamma_o|)$$

These values are used to define the *voltage standing-wave ratio* (VSWR), namely

$$\text{VSWR} = \frac{|V(x)|_{max}}{|V(x)|_{min}} = \frac{1 + |\Gamma_o|}{1 - |\Gamma_o|} \quad (1.3.11)$$

or

$$|\Gamma_o| = \frac{\text{VSWR} - 1}{\text{VSWR} + 1}$$

In a properly terminated or matched transmission line we obtain from (1.3.8), (1.3.10), and (1.3.11) that $\Gamma_0 = 0$, $Z_{IN}(d) = Z_o$, and VSWR = 1.

In a shorted transmission line $(Z_L = 0)$ it follows that $\Gamma_0 = -1$, VSWR = $\infty$, and the input impedance, called $Z_{sc}(d)$, is given by

$$Z_{sc}(d) = jZ_o \tan \beta d$$

In an open-circuited transmission line $(Z_L = \infty)$ it follows that $\Gamma_0 = 1$, VSWR = $\infty$, and the input impedance, called $Z_{oc}(d)$, is given by

$$Z_{oc}(d) = -jZ_o \cot \beta d$$

Another important case is the quarter-wave transmission line (also known as the quarter-wave transformer). With $d = \lambda/4$, (1.3.10) gives

$$Z_{IN}\left(\frac{\lambda}{4}\right) = \frac{Z_o^2}{Z_L} \qquad (1.3.12)$$

Equation (1.3.12) shows that in order to transform a real impedance $Z_L$ to another real impedance given by $Z_{IN}(\lambda/4)$, a line with characteristic impedance

$$Z_o = \sqrt{Z_{IN}\left(\frac{\lambda}{4}\right) Z_L}$$

can be used.

## 1.4 THE SCATTERING MATRIX AND THE CHAIN SCATTERING MATRIX

Introducing the notation

$$V^+(x) = Ae^{-\gamma x}$$

and

$$V^-(x) = Be^{\gamma x}$$

where $\gamma = j\beta$ for a lossless transmission line, we can write (1.3.3) and (1.3.4) in the form

$$V(x) = V^+(x) + V^-(x) \qquad (1.4.1)$$

and

$$I(x) = I^+(x) - I^-(x) = \frac{V^+(x)}{Z_o} - \frac{V^-(x)}{Z_o} \qquad (1.4.2)$$

Also, the reflection coefficient between the incident and reflected wave can be

## Sec. 1.4 The Scattering Matrix and the Chain Scattering Matrix

written as

$$\Gamma(x) = \frac{V^-(x)}{V^+(x)} \quad (1.4.3)$$

where $\Gamma_0 = \Gamma(0) = V^-(0)/V^+(0)$ is the load reflection coefficient.

Introducing the normalized notation

$$v(x) = \frac{V(x)}{\sqrt{Z_o}}$$

$$i(x) = \sqrt{Z_o}\, I(x)$$

$$a(x) = \frac{V^+(x)}{\sqrt{Z_o}}$$

and

$$b(x) = \frac{V^-(x)}{\sqrt{Z_o}}$$

we can write (1.4.1), (1.4.2), and (1.4.3) in the form

$$v(x) = a(x) + b(x)$$

$$i(x) = a(x) - b(x)$$

and

$$b(x) = \Gamma(x)a(x) \quad (1.4.4)$$

If instead of a one-port transmission line we have the two-port network shown in Fig. 1.4.1 with incident wave $a_1$ and reflected wave $b_1$ at port 1, and incident wave $a_2$ and reflected wave $b_2$ at port 2, we can generalize (1.4.4) and write

$$b_1 = S_{11}a_1 + S_{12}a_2$$

and

$$b_2 = S_{21}a_1 + S_{22}a_2$$

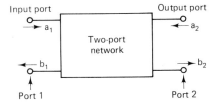

**Figure 1.4.1** Incident and reflected waves in a two-port network.

or in matrix form,

$$\begin{bmatrix} b_1 \\ b_2 \end{bmatrix} = \begin{bmatrix} S_{11} & S_{12} \\ S_{21} & S_{22} \end{bmatrix} \begin{bmatrix} a_1 \\ a_2 \end{bmatrix} \qquad (1.4.5)$$

Observe that $a_1$, $a_2$, $b_1$, and $b_2$ are values of the incident and reflected waves at the specific locations denoted as port 1 and port 2 in Fig. 1.4.1. The term $S_{11}a_1$ represents the contribution to the reflected wave $b_1$ due to the incident wave $a_1$ at port 1. Similarly, $S_{12}a_2$ represents the contribution to the reflected wave $b_1$ due to the incident wave $a_2$ at port 2, and so on. The parameters $S_{11}$, $S_{12}$, $S_{21}$, and $S_{22}$, which represent reflection and transmission coefficients, are called the *scattering parameters* of the two-port network. The matrix

$$[S] = \begin{bmatrix} S_{11} & S_{12} \\ S_{21} & S_{22} \end{bmatrix}$$

is called the *scattering matrix*.

The S parameters are seen to represent reflection or transmission coefficients. From (1.4.5), they are defined as follows:

$$S_{11} = \left. \frac{b_1}{a_1} \right|_{a_2 = 0} \qquad \text{(input reflection coefficient with output properly terminated)}$$

$$S_{21} = \left. \frac{b_2}{a_1} \right|_{a_2 = 0} \qquad \text{(forward transmission coefficient with output properly terminated)}$$

$$S_{22} = \left. \frac{b_2}{a_2} \right|_{a_1 = 0} \qquad \text{(output reflection coefficient with input properly terminated)}$$

$$S_{12} = \left. \frac{b_1}{a_2} \right|_{a_1 = 0} \qquad \text{(reverse transmission coefficient with input properly terminated)}$$

The advantage of using $S$ parameters is clear from their definitions. They are measured using a matched termination (i.e., making $a_1 = 0$ or $a_2 = 0$). For example, to measure $S_{11}$ we measure the ratio $b_1/a_1$ at the input port with the output port properly terminated, that is, with $a_2 = 0$. Terminating the output port with an impedance equal to the characteristic impedance of the transmission line produces $a_2 = 0$, because a traveling wave incident on the load will be totally absorbed and no energy will be returned to the output port. This situation is illustrated in Fig. 1.4.2.

Observe that the network output impedance $Z_{OUT}$ does not have to be matched to $Z_{o2}$. In fact, it is rare that $Z_{OUT} = Z_{o2}$, but with $Z_L = Z_{o2}$ the condition $a_2 = 0$ is satisfied. Similar considerations apply to measurements at the input port. Also, the characteristic impedances of the transmission lines are usually identical (i.e., $Z_{o1} = Z_{o2}$).

Using matched resistive terminations to measure the S parameters of a transistor has the advantage that the transistor does not oscillate. In contrast, if we were to use a short- or open-circuit test, the transistor could become unstable.

## Sec. 1.4  The Scattering Matrix and the Chain Scattering Matrix

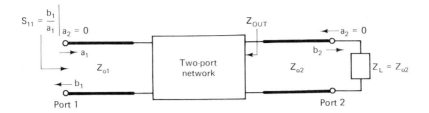

**Figure 1.4.2** Procedure for measuring $S_{11}$. The characteristic impedances of the transmission lines are $Z_{o1}$ and $Z_{o2}$.

The *chain scattering parameters*, also called the *scattering transfer parameters* or *T* parameters, are used when cascading networks. They are defined in such a way that the input waves $a_1$ and $b_1$ in Fig. 1.4.1 are the dependent variables and the output waves $a_2$ and $b_2$ are the independent variables. That is,

$$\begin{bmatrix} a_1 \\ b_1 \end{bmatrix} = \begin{bmatrix} T_{11} & T_{12} \\ T_{21} & T_{22} \end{bmatrix} \begin{bmatrix} b_2 \\ a_2 \end{bmatrix} \quad (1.4.6)$$

The relationship between the *S* and *T* parameters can be developed from (1.4.5) and (1.4.6), namely

$$\begin{bmatrix} T_{11} & T_{12} \\ T_{21} & T_{22} \end{bmatrix} = \begin{bmatrix} \dfrac{1}{S_{21}} & -\dfrac{S_{22}}{S_{21}} \\ \dfrac{S_{11}}{S_{21}} & S_{12} - \dfrac{S_{11}S_{22}}{S_{21}} \end{bmatrix} \quad (1.4.7)$$

and

$$\begin{bmatrix} S_{11} & S_{12} \\ S_{21} & S_{22} \end{bmatrix} = \begin{bmatrix} \dfrac{T_{21}}{T_{11}} & T_{22} - \dfrac{T_{21}T_{12}}{T_{11}} \\ \dfrac{1}{T_{11}} & -\dfrac{T_{12}}{T_{11}} \end{bmatrix} \quad (1.4.8)$$

The *T* parameters are useful in the analysis of cascade connections of two-port networks. Figure 1.4.3 shows that the output waves of the first network are identical to the input waves of the second network, namely

$$\begin{bmatrix} b_2 \\ a_2 \end{bmatrix} = \begin{bmatrix} a'_1 \\ b'_1 \end{bmatrix}$$

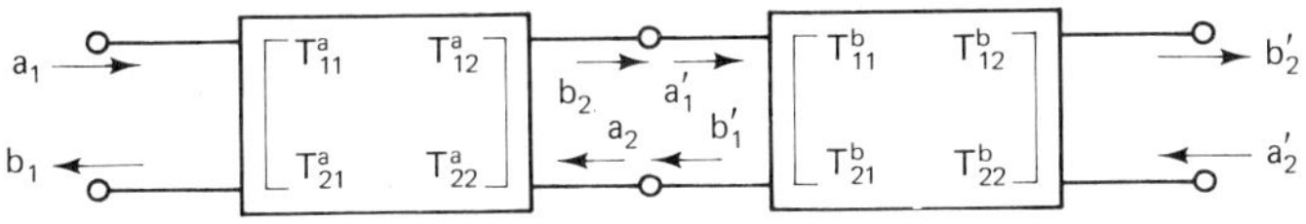

**Figure 1.4.3** Cascade connection of two-port networks.

Therefore, the chain scattering matrix of the cascade connection can be written in terms of the individual chain scattering matrix as follows:

$$\begin{bmatrix} a_1 \\ b_1 \end{bmatrix} = \begin{bmatrix} T^a_{11} & T^a_{12} \\ T^a_{21} & T^a_{22} \end{bmatrix} \begin{bmatrix} b_2 \\ a_2 \end{bmatrix}$$

$$\begin{bmatrix} a'_1 \\ b'_1 \end{bmatrix} = \begin{bmatrix} T^b_{11} & T^b_{12} \\ T^b_{21} & T^b_{22} \end{bmatrix} \begin{bmatrix} b'_2 \\ a'_2 \end{bmatrix}$$

and

$$\begin{bmatrix} a_1 \\ b_1 \end{bmatrix} = \begin{bmatrix} T^a_{11} & T^a_{12} \\ T^a_{21} & T^a_{22} \end{bmatrix} \begin{bmatrix} T^b_{11} & T^b_{12} \\ T^b_{21} & T^b_{22} \end{bmatrix} \begin{bmatrix} b'_2 \\ a'_2 \end{bmatrix} \quad (1.4.9)$$

Equation (1.4.9) is useful in the analysis and design of microwave amplifiers using computer-aided design techniques.

## 1.5 SHIFTING REFERENCE PLANES

In practice we often need to attach transmission lines to the two-port network. Since the $S$ parameters are measured using traveling waves, we need to specify the positions where the measurements are made. The positions are called *reference planes*. For example, in Fig. 1.5.1 we can measure the $S$ parameters at the reference planes located at port $1'$ and port $2'$ and relate them to the $S$ parameters at port 1 and port 2 of the two-port network.

At the reference planes at port 1 and port 2 in Fig. 1.5.1, we write the scattering matrix as

$$\begin{bmatrix} b_1 \\ b_2 \end{bmatrix} = \begin{bmatrix} S_{11} & S_{12} \\ S_{21} & S_{22} \end{bmatrix} \begin{bmatrix} a_1 \\ a_2 \end{bmatrix} \quad (1.5.1)$$

and at port $1'$ and port $2'$ as

$$\begin{bmatrix} b'_1 \\ b'_2 \end{bmatrix} = \begin{bmatrix} S'_{11} & S'_{12} \\ S'_{21} & S'_{22} \end{bmatrix} \begin{bmatrix} a'_1 \\ a'_2 \end{bmatrix} \quad (1.5.2)$$

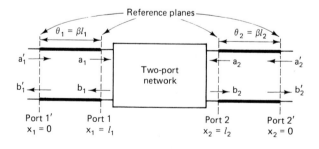

**Figure 1.5.1** Model for shifting reference planes.

Sec. 1.6    Properties of Scattering Parameters

The angles $\theta_1$ and $\theta_2$ are the electrical lengths of the transmission lines between the primed and unprimed reference planes.

From our knowledge of traveling waves on a lossless transmission line we can write

$$b_1 = b'_1 e^{j\theta_1}$$

$$a_1 = a'_1 e^{-j\theta_1}$$

$$b_2 = b'_2 e^{j\theta_2}$$

and

$$a_2 = a'_2 e^{-j\theta_2}$$

where the factor $e^{\pm j\theta}$ accounts for the phase difference of the waves at the different reference planes. Substituting the previous relations into (1.5.1) gives

$$\begin{bmatrix} b'_1 \\ b'_2 \end{bmatrix} = \begin{bmatrix} S_{11} e^{-j2\theta_1} & S_{12} e^{-j(\theta_1+\theta_2)} \\ S_{21} e^{-j(\theta_1+\theta_2)} & S_{22} e^{-j2\theta_2} \end{bmatrix} \begin{bmatrix} a'_1 \\ a'_2 \end{bmatrix} \quad (1.5.3)$$

Comparing (1.5.3) with (1.5.2) gives the relations

$$\begin{bmatrix} S'_{11} & S'_{12} \\ S'_{21} & S'_{22} \end{bmatrix} = \begin{bmatrix} S_{11} e^{-j2\theta_1} & S_{12} e^{-j(\theta_1+\theta_2)} \\ S_{21} e^{-j(\theta_1+\theta_2)} & S_{22} e^{-j2\theta_2} \end{bmatrix} \quad (1.5.4)$$

and

$$\begin{bmatrix} S_{11} & S_{12} \\ S_{21} & S_{22} \end{bmatrix} = \begin{bmatrix} S'_{11} e^{j2\theta_1} & S'_{12} e^{j(\theta_1+\theta_2)} \\ S'_{21} e^{j(\theta_1+\theta_2)} & S'_{22} e^{j2\theta_2} \end{bmatrix} \quad (1.5.5)$$

Equations (1.5.4) and (1.5.5) provide the relationship between $S$ parameters at two sets of reference planes.

## 1.6 PROPERTIES OF SCATTERING PARAMETERS

Consider the two-port network shown in Fig. 1.6.1, where the transmission lines are assumed to be lossless and the characteristic impedances are *real*. This is the typical situation at microwave frequencies where 50-$\Omega$ transmission lines and 50-$\Omega$ terminations are commonly used. From (1.4.1) and (1.4.2) the

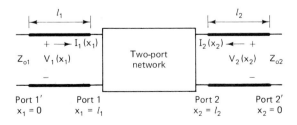

**Figure 1.6.1**  Two-port network.

voltages and currents in the two-port network can be written as

$$V_i(x_i) = V_i^+(x_i) + V_i^-(x_i) \tag{1.6.1}$$

and

$$I_i(x_i) = I_i^+(x_i) - I_i^-(x_i) = \frac{V_i^+(x_i)}{Z_{oi}} - \frac{V_i^-(x_i)}{Z_{oi}} \tag{1.6.2}$$

where $i = 1$ or 2, and the voltages and currents are assumed to be scaled to root-mean-square (rms) values.

From (1.6.1) and (1.6.2) we can express the normalized incident and reflected voltages at the $i$th port in the form

$$a_i(x_i) = \frac{V_i^+(x_i)}{\sqrt{Z_{oi}}} = \sqrt{Z_{oi}}\, I_i^+(x_i) = \frac{1}{2\sqrt{Z_{oi}}}[V_i(x_i) + Z_{oi} I_i(x_i)] \tag{1.6.3}$$

and

$$b_i(x_i) = \frac{V_i^-(x_i)}{\sqrt{Z_{oi}}} = \sqrt{Z_{oi}}\, I_i^-(x_i) = \frac{1}{2\sqrt{Z_{oi}}}[V_i(x_i) - Z_{oi} I_i(x_i)] \tag{1.6.4}$$

The average power associated with the incident waves on the primed $i$th port (i.e., at $x_1 = 0$ and $x_2 = 0$) is

$$P_i^+(0) = \frac{|V_i^+(0)|^2}{Z_{oi}} = a_i(0)a_i^*(0) = |a_i(0)|^2 \tag{1.6.5}$$

and the average reflected power is

$$P_i^-(0) = \frac{|V_i^-(0)|^2}{Z_{oi}} = b_i(0)b_i^*(0) = |b_i(0)|^2 \tag{1.6.6}$$

Since the line is lossless [i.e., $P_i^+(0) = P_i^+(l_i)$ and $P_i^-(0) = P_i^-(l_i)$], (1.6.5) and (1.6.6) show that the quantities $|a_i(x)|^2$ and $|b_i(x)|^2$ represent the power associated with the incident and reflected waves, respectively.

Now consider the network in Fig. 1.6.2, in which port 1' is excited by the generator $E_1 \,\underline{/0°}$ V rms having impedance $Z_{o1}$, and port 2' is terminated in its

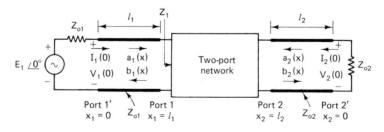

**Figure 1.6.2** Two-port network excited by the generator $E_1 \,\underline{/0°}$ V rms with internal impedance $Z_{o1}$ and terminated in the impedance $Z_{o2}$.

Sec. 1.6   Properties of Scattering Parameters

normalizing impedance $Z_{o2}$ (i.e., matched impedance). Again, we point out that in practice $Z_{o1}$ and $Z_{o2}$ are pure resistors (usually 50 Ω). Since port 2 is terminated in its normalizing impedance, we can write

$$V_2(0) = -Z_{o2} I_2(0) \tag{1.6.7}$$

and from (1.6.3) it follows that $a_2(0) = 0$.

At $x_1 = 0$, we have

$$V_1(0) = E_1 - Z_{o1} I_1(0) \tag{1.6.8}$$

Substituting (1.6.8) into (1.6.3) gives

$$a_1(0) = \frac{E_1}{2\sqrt{Z_{o1}}}$$

or

$$|a_1(0)|^2 = \frac{|E_1|^2}{4Z_{o1}} \tag{1.6.9}$$

Equation (1.6.9) shows that $|a_1(0)|^2$ represents the power available from the source $E_1$ with internal resistance $Z_{o1}$. We call this quantity $P_{\text{AVS}}$. Since the line is lossless ($|a_1(0)|^2 = |a_1(l_1)|^2$) $|a_1(0)|^2$ represents the power available at port 1. The power available from the source $E_1$ is independent of the input impedance of the two-port network.

Substituting (1.6.8) into (1.6.9) gives

$$|a_1(0)|^2 = \frac{E_1 E_1^*}{4Z_{o1}} = \frac{[V_1(0) + Z_{o1} I_1(0)][V_1(0) + Z_{o1} I_1(0)]^*}{4Z_{o1}}$$

$$= \frac{1}{4Z_{o1}} [|V_1(0)|^2 + Z_{o1} I_1(0) V_1^*(0) + Z_{o1} V_1(0) I_1^*(0)$$

$$+ Z_{o1}^2 |I_1(0)|^2] \tag{1.6.10}$$

Similarly, from (1.6.4) we obtain

$$|b_1(0)|^2 = \frac{1}{4Z_{o1}} [|V_1(0)|^2 - Z_{o1} I_1(0) V_1^*(0) - Z_{o1} I_1^*(0) V_1(0) + Z_{o1}^2 |I_1(0)|^2] \tag{1.6.11}$$

Subtracting (1.6.11) from (1.6.10) gives

$$|a_1(0)|^2 - |b_1(0)|^2 = \tfrac{1}{2}[I_1(0) V_1^*(0) + I_1^*(0) V_1(0)]$$

$$= \text{Re}\,[I_1(0) V_1^*(0)]$$

which represents the power delivered to port 1′, or to port 1 since the line is lossless. We call this quantity $P_1$ [i.e., $P_1 = |a_1(0)|^2 - |b_1(0)|^2$]. Therefore, it follows that

$$|b_1(0)|^2 = P_{\text{AVS}} - P_1 \tag{1.6.12}$$

The quantity $|b_1(0)|^2$ represents the reflected power from port 1 or port 1'.
From (1.6.7) and (1.6.4) we obtain

$$b_2(0) = -\sqrt{Z_{o2}}\, I_2(0)$$

Therefore,

$$|b_2(0)|^2 = Z_{o2}|I_2(0)|^2$$

represents the power delivered to the load $Z_{o2}$, called $P_L$ [i.e., $P_L = |b_2(0)|^2$].

Equations (1.6.9) and (1.6.12) show that the generator sends the available power $|a_1(0)|^2$ toward the input port 1. This power is independent of the input impedance $Z_1$. If the input impedance $Z_1$ is matched to the transmission line (i.e., if $Z_1 = Z_{o1}$), then the reflected power is zero. However, if $Z_1 \neq Z_{o1}$, part of the incident power $|a_1(0)|^2$ is reflected back to the generator. The reflected power is given by $|b_1(0)|^2$ and the net power delivered to port 1 is $|a_1(0)|^2 - |b_1(0)|^2$.

In order to calculate the $S$ parameters of the two-port network at the unprimed reference plane (i.e., at ports 1 and 2) we replace the network in Fig. 1.6.2 by the equivalent network shown in Fig. 1.6.3. The equivalent network was obtained by finding the Thévenin's equivalent at ports 1 and 2. The Thévenin's voltage is called $E_{1,TH}$.

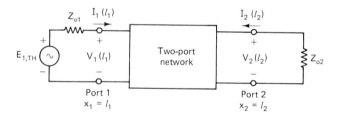

**Figure 1.6.3** Two-port network with Thévenin's equivalent at ports 1 and 2.

The scattering parameter $S_{11}$ is given by

$$S_{11} = \left.\frac{b_1(l_1)}{a_1(l_1)}\right|_{a_2(l_2)=0} = \left.\frac{V_1^-(l_1)}{V_1^+(l_1)}\right|_{V_2^+(l_2)=0} \tag{1.6.13}$$

which from (1.3.8) it can be expressed as

$$S_{11} = \frac{Z_1 - Z_{o1}}{Z_1 + Z_{o1}} \tag{1.6.14}$$

Equation (1.6.13) or (1.6.14) shows that $S_{11}$ is the reflection coefficient of port 1 with port 2 terminated in its normalizing impedance $Z_{o2}$ (i.e., $a_2 = 0$).

If we consider the quantity $|S_{11}|^2$ we find from (1.6.9) and (1.6.12) that

$$|S_{11}|^2 = \left.\frac{|b_1(l_1)|^2}{|a_1(l_1)|^2}\right|_{a_2(l_2)=0} = \frac{P_{AVS} - P_1}{P_{AVS}} \tag{1.6.15}$$

Equation (1.6.15) shows that $|S_{11}|^2$ represents the ratio of the power reflected from port 1 to the power available at port 1. If $|S_{11}| > 1$, the power reflected is larger than the power available at port 1. Therefore, in this case port 1 acts as a source of power and oscillations can occur.

The evaluation of $S_{21}$ is as follows:

$$S_{21} = \left.\frac{b_2(l_2)}{a_1(l_1)}\right|_{a_2(l_2)=0} = \left.\frac{\sqrt{Z_{o2}}\, I_2^-(l_2)}{\sqrt{Z_{o1}}\, I_1^+(l_1)}\right|_{I_2^-(l_2)=0}$$

$$= \left.\frac{-\sqrt{Z_{o2}}\, I_2(l_2)}{\sqrt{Z_{o1}}\, I_1^+(l_1)}\right|_{I_2^-(l_2)=0} \quad (1.6.16)$$

because $I_2(l_2) = -I_2^-(l_2)$ [i.e., $a_2(l_2) = I_2^+(l_2) = 0$]. Since

$$I_1^+(l_1) = \frac{E_{1,\text{TH}}}{2Z_{o1}}$$

and

$$V_2(l_2) = -Z_{o2}\, I_2(l_2)$$

we can write (1.6.16) in the form

$$S_{21} = \frac{2\sqrt{Z_{o1}}}{\sqrt{Z_{o2}}} \frac{V_2(l_2)}{E_{1,\text{TH}}} \quad (1.6.17)$$

Equation (1.6.17) shows that $S_{21}$ represents a forward voltage transmission coefficient from port 1 to port 2.

The analysis of $|S_{21}|^2$ gives

$$|S_{21}|^2 = \frac{|I_2(l_2)|^2 Z_{o2}}{|E_{1,\text{TH}}|^2/4Z_{o1}} \quad (1.6.18)$$

Equation (1.6.18) shows that $|S_{21}|^2$ represents the ratio of the power delivered to the load $Z_{o2}$ (i.e., $P_L$) to the power available from the source (i.e., $P_{\text{AVS}}$). The ratio $P_L/P_{\text{AVS}}$ is known as the *transducer power gain*.

If we analyze the network shown in Fig. 1.6.4 in which the excitation $E_2 \underline{/0°}$ V rms is placed in port 2′, and port 1′ is terminated in its normalizing impedance $Z_{o1}$, we find that

$$S_{22} = \left.\frac{b_2(l_2)}{a_2(l_2)}\right|_{a_1(l_1)=0} = \frac{Z_2 - Z_{o2}}{Z_2 + Z_{o2}} \quad (1.6.19)$$

and

$$S_{12} = \left.\frac{b_1(l_1)}{a_2(l_2)}\right|_{a_1(l_1)=0} = \frac{2\sqrt{Z_{o2}}}{\sqrt{Z_{o1}}} \frac{V_1(l_1)}{E_{2,\text{TH}}}$$

where $E_{2,\text{TH}}$ is the Thévenin's voltage at port 2. Equation (1.6.19) shows that $S_{22}$ is the reflection coefficient of port 2 with port 1 terminated in its normal-

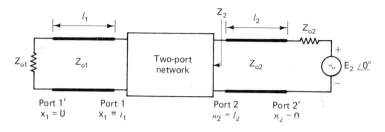

**Figure 1.6.4** Two-port network excited by the generator $E_2 \underline{/0°}$ V rms with internal impedance $Z_{o2}$ and terminated in the impedance $Z_{o1}$.

izing impedance $Z_{o1}$ (i.e., $a_1 = 0$), and $S_{12}$ represents a reverse voltage transmission coefficient from port 2 to port 1.

The quantity $|S_{22}|^2$ represents the ratio of the power reflected from port 2 to the power available at port 2. If $|S_{22}| > 1$, the power reflected is larger than the power available at port 2 and oscillation can occur. The quantity $|S_{12}|^2$ represents a reverse transducer power gain. In fact,

$$|S_{12}|^2 = \frac{|I_1(l_1)|^2 Z_{o1}}{|E_{2,\text{TH}}|^2/4Z_{o2}}$$

**Example 1.6.1**

Evaluate the S parameters of a series impedance Z and a shunt admittance Y.

**Solution.** The two-port network of a series impedance is shown in Fig. 1.6.5a, and the network excited by a source and terminated in its normalizing impedance $Z_o$ is shown in Fig. 1.6.5b.

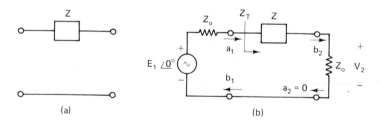

(a)    (b)

**Figure 1.6.5** Two-port network of a series impedance Z.

From (1.6.14) we find that

$$S_{11} = \frac{b_1}{a_1}\bigg|_{a_2=0} = \frac{Z_T - Z_o}{Z_T + Z_o}$$

where $Z_T = Z + Z_o$. Therefore,

$$S_{11} = \frac{Z}{Z + 2Z_o}$$

Since

$$V_2 = \frac{E_1 Z_o}{Z + 2Z_o}$$

we find from (1.6.17) that

$$S_{21} = \frac{2Z_o}{Z + 2Z_o}$$

From symmetry, we observe that $S_{22} = S_{11}$ and $S_{12} = S_{21}$.

The two-port network of a shunt admittance is shown in Fig. 1.6.6a, and the terminated network is shown in Fig. 1.6.6b. In this case

$$Z_T = \frac{Z_o}{1 + Z_o Y}$$

and from (1.6.14)

$$S_{11} = \frac{Z_T - Z_o}{Z_T + Z_o} = \frac{-Z_o Y}{2 + Z_o Y}$$

Since

$$V_2 = \frac{E_1}{2 + Z_o Y}$$

we obtain from (1.6.17)

$$S_{21} = \frac{2}{2 + Z_o Y}$$

Again, from symmetry we observe that $S_{22} = S_{11}$ and $S_{12} = S_{21}$.

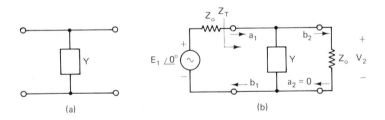

**Figure 1.6.6** Two-port network of a shunt admittance $Y$.

## 1.7. GENERALIZED SCATTERING PARAMETERS

The typical situation that occurs at microwave frequencies, where 50-Ω transmission lines and 50-Ω terminations are commonly used to measure the $S$ parameters of a two-port network, was analyzed in the preceding section. In this section we consider the case of the $n$-port network shown in Fig. 1.7.1. The

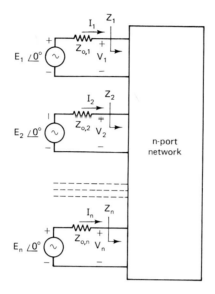

**Figure 1.7.1** An $n$-port network.

generators $E_i \underline{/0°}$ and the impedances $Z_{o,i}$ represent the Thévenin equivalent circuits seen by the $i$th ports. The impedances $Z_{o,i}$ (also called normalizing impedances) are assumed to have a positive real part.

In the $n$-port network the normalized incident and reflected waves are defined as

$$[a] = \tfrac{1}{2}[R_{o,i}^{-1/2}]([V] + [Z_{o,i}][I]) \tag{1.7.1}$$

and

$$[b] = \tfrac{1}{2}[R_{o,i}^{-1/2}]([V] - [Z_{o,i}]^*[I]) \tag{1.7.2}$$

where

$$[Z_{o,i}] = \begin{bmatrix} Z_{o,1} & 0 & \cdots & 0 \\ 0 & Z_{o,2} & \cdots & 0 \\ \vdots & & & \vdots \\ 0 & 0 & \cdots & Z_{o,n} \end{bmatrix}$$

and

$$[R_{o,i}^{-1/2}] = \begin{bmatrix} (\text{Re } Z_{o,1})^{-1/2} & 0 & \cdots & 0 \\ 0 & (\text{Re } Z_{o,2})^{-1/2} & \cdots & 0 \\ \vdots & & & \vdots \\ 0 & \cdots & \cdots & (\text{Re } Z_{o,n})^{-1/2} \end{bmatrix}$$

$[a]$, $[b]$, $[V]$, and $[I]$ are column matrices.

The definitions (1.7.1) and (1.7.2) are generalizations of (1.6.3) and (1.6.4).

Sec. 1.7  Generalized Scattering Parameters

In fact, at port 1 we can write

$$a_1 = \tfrac{1}{2} R_{o,1}^{-1/2}(V_1 + Z_{o,1} I_1) \quad (1.7.3)$$

and

$$b_1 = \tfrac{1}{2} R_{o,1}^{-1/2}(V_1 - Z_{o,1}^* I_1) \quad (1.7.4)$$

and the relation between $V_1$ and $I_1$ is

$$V_1 = E_1 - Z_{o,1} I_1 \quad (1.7.5)$$

Substituting (1.7.5) into (1.7.3) gives

$$a_1 = \frac{E_1}{2 R_{o,1}^{1/2}}$$

or

$$|a_1|^2 = \frac{|E_1|^2}{4 R_{o,1}} \quad (1.7.6)$$

which is recognized as the power available from $E_1$ at port 1. Also, from (1.7.3) and (1.7.4) we obtain

$$|a_1|^2 - |b_1|^2 = \frac{1}{4 R_{o,1}} (V_1 + Z_{o,1} I_1)(V_1^* + Z_{o,1}^* I_1^*)$$

$$- \frac{1}{4 R_{o,1}} (V_1 - Z_{o,1}^* I_1)(V_1^* - Z_{o,1} I_1^*)$$

$$= \text{Re}\,(V_1^* I_1) \quad (1.7.7)$$

which is recognized as the power delivered to port 1.

Equations (1.7.4), (1.7.6), and (1.7.7) show that the generator $E_1$ sends the available power $|a_1|^2$ toward port 1. When port 1 is matched [i.e., when $Z_1 = V_1/I_1 = (Z_{o,1})^*$] the power $|a_1|^2$ is completely absorbed by $Z_1$. When port 1 is not matched, the power absorbed by the port 1 is $|a_1|^2 - |b_1|^2$, where $|b_1|^2$ is the reflected power.

Solving (1.7.1) and (1.7.2) for $[V]$ and $[I]$ results in

$$[V] = [V^+] + [V^-]$$
$$[I] = [I^+] - [I^-]$$

where

$$[V^+] = [Z_{o,i}^*][I^+]$$
$$[V^-] = [Z_{o,i}][I^-]$$
$$[a] = [R_{o,i}^{1/2}][I^+]$$

and
$$[b] = [R_{o,i}^{1/2}][I^-]$$

The generalized scattering matrix of the $n$-port network is defined as
$$[b] = [S][a] \qquad (1.7.8)$$

The definition (1.7.8) of the scattering matrix shows that different normalizing impedances $Z_{o,i}$ produce different values of the generalized scattering parameters. Therefore, the generalized scattering parameters are defined in terms of specific normalizing impedances. If the normalizing impedances are pure resistances, the results in Section 1.6 for the two-port network follow.

From (1.7.8), the elements of $[S]$ are given by
$$S_{ii} = \left.\frac{b_i}{a_i}\right|_{a_k = 0,\, k \neq i\, (k = 0, 1, \ldots, n)}$$

and
$$S_{ki} = \left.\frac{b_k}{a_i}\right|_{a_k = 0,\, k \neq i\, (k = 0, 1, \ldots, n)}$$

$S_{ii}$ is recognized as the input reflection coefficient at port 1 with all other ports matched (i.e., $a_k = 0$ when $V_k = -Z_{o,k} I_k$, $k \neq i$). Observing that
$$V_i = Z_i I_i$$

then from the $i$th equation in (1.7.1) and (1.7.2), we obtain
$$S_{ii} = \left.\frac{b_i}{a_i}\right|_{a_k = 0,\, k \neq i} = \frac{V_i - Z_{o,i}^* I_i}{V_i + Z_{o,i} I_i} = \frac{Z_i - Z_{o,i}^*}{Z_i + Z_{o,i}}$$

The quantity $|S_{ki}|^2$ can be shown to be the transducer power gain from port $i$ to port $k$ with $a_k = 0$, $k \neq i$.

## 1.8 TWO-PORT NETWORK PARAMETERS CONVERSIONS

At a given frequency a two-port network can be described in terms of several parameters. Therefore, it is desirable to have relations to convert from one set of parameters to another. For example, the $z$ parameters of a two-port network are defined by
$$[V] = [z][I] \qquad (1.8.1)$$

where
$$[V] = \begin{bmatrix} v_1 \\ v_2 \end{bmatrix}$$

$$[I] = \begin{bmatrix} i_1 \\ i_2 \end{bmatrix}$$

and

$$[z] = \begin{bmatrix} z_{11} & z_{12} \\ z_{21} & z_{22} \end{bmatrix}$$

In terms of incident and reflected waves, we obtain from (1.8.1)

$$[V^+] + [V^-] = [z]([I^+] - [I^-])$$

or

$$([z] + [Z_o])[I^-] = ([z] - [Z_o])[I^+]$$

where $Z_o$ is assumed to be real and

$$[Z_o] = \begin{bmatrix} Z_o & 0 \\ 0 & Z_o \end{bmatrix}$$

Therefore, the scattering matrix [i.e., (1.7.8)] in terms of $z$ parameters is given by

$$[S] = \frac{[I^-]}{[I^+]} = ([z] + [Z_o])^{-1}([z] - [Z_o]) \qquad (1.8.2)$$

and solving for $[z]$, we obtain

$$[z] = [Z_o]([1] + [S])([1] - [S])^{-1} \qquad (1.8.3)$$

where $[1]$ is the unit diagonal matrix. Equations (1.8.2) and (1.8.3) give the conversion relations between the $S$ and $z$ parameters. These conversions, as well as others among the $z$, $y$, $h$, $ABCD$, and $S$ parameters, are tabulated in Fig. 1.8.1.

## 1.9 SCATTERING PARAMETERS OF TRANSISTORS

The $S$ parameters of microwave transistors are usually available for the transistor in chip and packaged form. Transistors in chip form are used when the best performance in gain, bandwidth, and noise is desired. Packaged transistors are very popular because they come in sealed enclosures and are easy to work with. The parasitic elements introduced by the package produce a degradation in the transistor ac performance.

Manufacturers usually measure and provide common-emitter or common-source $S$ parameters of transistors as a function of frequency at a given dc bias. Since the minimum noise figure, linear output power, and maximum gain require different dc bias settings, the manufacturers usually provide two or three sets of $S$ parameters.

Conversions from common-emitter to common-base $S$ parameters can be done using the conversion relations in Fig. 1.8.1. For example, to convert from common-emitter to common-base $S$ parameters, we first convert the common-

|   | s | z | y | h | ABCD |
|---|---|---|---|---|---|
| **s** | $S_{11}\ \ S_{12}$<br>$S_{21}\ \ S_{22}$ | $S_{11} = \dfrac{(z'_{11} - 1)(z'_{22} + 1) - z'_{12}z'_{21}}{\Delta_1}$<br>$S_{12} = \dfrac{2z'_{12}}{\Delta_1}$<br>$S_{21} = \dfrac{2z'_{21}}{\Delta_1}$<br>$S_{22} = \dfrac{(z'_{11}+1)(z'_{22}-1) - z'_{12}z'_{21}}{\Delta_1}$ | $S_{11} = \dfrac{(1-y'_{11})(1+y'_{22}) + y'_{12}y'_{21}}{\Delta_2}$<br>$S_{12} = \dfrac{-2y'_{12}}{\Delta_2}$<br>$S_{21} = \dfrac{-2y'_{21}}{\Delta_2}$<br>$S_{22} = \dfrac{(1+y'_{11})(1-y'_{22}) + y'_{12}y'_{21}}{\Delta_2}$ | $S_{11} = \dfrac{(h'_{11} - 1)(h'_{22} + 1) - h'_{12}h'_{21}}{\Delta_3}$<br>$S_{12} = \dfrac{2h'_{12}}{\Delta_3}$<br>$S_{21} = \dfrac{-2h'_{21}}{\Delta_3}$<br>$S_{22} = \dfrac{(1+h'_{11})(1-h'_{22}) + h'_{12}h'_{21}}{\Delta_3}$ | $S_{11} = \dfrac{A' + B' - C' - D'}{\Delta_4}\ \ \ S_{12} = \dfrac{2(A'D' - B'C')}{\Delta_4}$<br><br>$S_{21} = \dfrac{2}{\Delta_4}\ \ \ S_{22} = \dfrac{-A'+B'-C'+D'}{\Delta_4}$ |
| **z** | $z'_{11} = \dfrac{(1+S_{11})(1-S_{22}) + S_{12}S_{21}}{\Delta_5}$<br>$z'_{12} = \dfrac{2S_{12}}{\Delta_5}$<br>$z'_{21} = \dfrac{2S_{21}}{\Delta_5}$<br>$z'_{22} = \dfrac{(1-S_{11})(1+S_{22}) + S_{12}S_{21}}{\Delta_5}$ | $z_{11}\ \ z_{12}$<br>$z_{21}\ \ z_{22}$ | $\dfrac{y_{22}}{\lvert y\rvert}\ \ \dfrac{-y_{12}}{\lvert y\rvert}$<br>$\dfrac{-y_{21}}{\lvert y\rvert}\ \ \dfrac{y_{11}}{\lvert y\rvert}$ | $\dfrac{\lvert h\rvert}{h_{22}}\ \ \dfrac{h_{12}}{h_{22}}$<br>$\dfrac{-h_{21}}{h_{22}}\ \ \dfrac{1}{h_{22}}$ | $\dfrac{A}{C}\ \ \dfrac{\Delta_8}{C}$<br>$\dfrac{1}{C}\ \ \dfrac{D}{C}$ |
| **y** | $y'_{11} = \dfrac{(1-S_{11})(1+S_{22}) + S_{12}S_{21}}{\Delta_6}$<br>$y'_{12} = \dfrac{-2S_{12}}{\Delta_6}$<br>$y'_{21} = \dfrac{-2S_{21}}{\Delta_6}$<br>$y'_{22} = \dfrac{(1+S_{11})(1-S_{22}) + S_{12}S_{21}}{\Delta_6}$ | $\dfrac{z_{22}}{\lvert z\rvert}\ \ \dfrac{-z_{12}}{\lvert z\rvert}$<br>$\dfrac{-z_{21}}{\lvert z\rvert}\ \ \dfrac{z_{11}}{\lvert z\rvert}$ | $y_{11}\ \ y_{12}$<br>$y_{21}\ \ y_{22}$ | $\dfrac{1}{h_{11}}\ \ \dfrac{-h_{12}}{h_{11}}$<br>$\dfrac{h_{21}}{h_{11}}\ \ \dfrac{\lvert h\rvert}{h_{11}}$ | $\dfrac{D}{B}\ \ \dfrac{-\Delta_8}{B}$<br>$\dfrac{-1}{B}\ \ \dfrac{A}{B}$ |

|  | z | y | h | ABCD |
|---|---|---|---|---|
| h | $h'_{11} = \dfrac{(1+S_{11})(1+S_{22}) - S_{12}S_{21}}{\Delta_7}$<br><br>$h'_{12} = \dfrac{2S_{12}}{\Delta_7}$<br><br>$h'_{21} = \dfrac{-2S_{21}}{\Delta_7}$<br><br>$h'_{22} = \dfrac{(1-S_{22})(1-S_{11}) - S_{12}S_{21}}{\Delta_7}$ | $\dfrac{\lvert z\rvert}{z_{22}} \quad \dfrac{z_{12}}{z_{22}}$<br><br>$\dfrac{-z_{21}}{z_{22}} \quad \dfrac{1}{z_{22}}$ | $\dfrac{1}{y_{11}} \quad \dfrac{-y_{12}}{y_{11}}$<br><br>$\dfrac{y_{21}}{y_{11}} \quad \dfrac{\lvert y\rvert}{y_{11}}$ | $h_{11} \quad h_{12}$<br><br>$h_{21} \quad h_{22}$ | $\dfrac{B}{D} \quad \dfrac{\Delta_8}{D}$<br><br>$\dfrac{-1}{D} \quad \dfrac{C}{D}$ |
| ABCD | $A' = \dfrac{(1+S_{11})(1-S_{22}) + S_{12}S_{21}}{2S_{21}}$<br><br>$B' = \dfrac{(1+S_{11})(1+S_{22}) - S_{12}S_{21}}{2S_{21}}$<br><br>$C' = \dfrac{(1-S_{11})(1-S_{22}) - S_{12}S_{21}}{2S_{21}}$<br><br>$D' = \dfrac{(1-S_{11})(1+S_{22}) + S_{12}S_{21}}{2S_{21}}$ | $\dfrac{z_{11}}{z_{21}} \quad \dfrac{\lvert z\rvert}{z_{21}}$<br><br>$\dfrac{1}{z_{21}} \quad \dfrac{z_{22}}{z_{21}}$ | $\dfrac{-y_{22}}{y_{21}} \quad \dfrac{-1}{y_{21}}$<br><br>$\dfrac{-\lvert y\rvert}{y_{21}} \quad \dfrac{-y_{11}}{y_{21}}$ | $\dfrac{-\lvert h\rvert}{h_{21}} \quad \dfrac{-h_{11}}{h_{21}}$<br><br>$\dfrac{-h_{22}}{h_{21}} \quad \dfrac{-1}{h_{21}}$ | $A \quad B$<br><br>$C \quad D$ |

$\Delta_1 = (z'_{11} + 1)(z'_{22} + 1) - z'_{12}z'_{21}$
$\Delta_2 = (1 + y'_{11})(1 + y'_{22}) - y'_{12}y'_{21}$
$\Delta_3 = (h'_{11} + 1)(h'_{22} + 1) - h'_{12}h'_{21}$
$\Delta_4 = A' + B' + C' + D'$
$\Delta_5 = (1 - S_{11})(1 - S_{22}) - S_{12}S_{21}$
$\Delta_6 = (1 + S_{11})(1 + S_{22}) - S_{12}S_{21}$
$\Delta_7 = (1 - S_{11})(1 + S_{22}) + S_{12}S_{21}$
$\Delta_8 = AD - BC$

$z'_{11} = z_{11}/Z_o, \quad z'_{12} = z_{12}/Z_o, \quad z'_{21} = z_{21}/Z_o, \quad z'_{22} = z_{22}/Z_o$
$y'_{11} = y_{11}Z_o, \quad y'_{12} = y_{12}Z_o, \quad y'_{21} = y_{21}Z_o, \quad y'_{22} = y_{22}Z_o$
$h'_{11} = h_{11}/Z_o, \quad h'_{12} = h_{12}, \quad h'_{21} = h_{21}, \quad h'_{22} = h_{22}Z_o$
$A' = A, \quad B' = B/Z_o, \quad C' = CZ_o, \quad D' = D$
$\lvert z\rvert = z_{11}z_{22} - z_{12}z_{21}$
$\lvert y\rvert = y_{11}y_{22} - y_{12}y_{21}$
$\lvert h\rvert = h_{11}h_{22} - h_{12}h_{21}$

(a)

*Conversion Between Common-Base, Common-Emitter, and Common-Collector y Parameters*

$$y_{11,e} = y_{11,b} + y_{12,b} + y_{21,b} + y_{22,b} = y_{11,c}$$
$$y_{12,e} = -(y_{12,b} + y_{22,b}) = -(y_{11,c} + y_{12,c})$$
$$y_{21,e} = -(y_{21,b} + y_{22,b}) = -(y_{11,c} + y_{21,c})$$
$$y_{22,e} = y_{22,b} = y_{11,c} + y_{12,c} + y_{21,c} + y_{22,c}$$

$$y_{11,b} = y_{11,e} + y_{12,e} + y_{21,e} + y_{22,e} = y_{22,c}$$
$$y_{12,b} = -(y_{12,e} + y_{22,e}) = -(y_{21,c} + y_{22,c})$$
$$y_{21,b} = -(y_{21,e} + y_{22,e}) = -(y_{12,c} + y_{22,c})$$
$$y_{22,b} = y_{22,e} = y_{11,c} + y_{12,c} + y_{21,c} + y_{22,c}$$

$$y_{11,c} = y_{11,e} = y_{11,b} + y_{12,b} + y_{21,b} + y_{22,b}$$
$$y_{11,c} = -(y_{11,e} + y_{12,e}) = -(y_{11,b} + y_{21,b})$$
$$y_{21,c} = -(y_{11,e} + y_{21,e}) = -(y_{11,b} + y_{12,b})$$
$$y_{22,c} = y_{11,e} + y_{12,e} + y_{21,e} + y_{22,e} = y_{11,b}$$

(b)

**Figure 1.8.1** (a) Conversions among the $z$, $y$, $h$, $ABCD$, and $S$ parameters; (b) conversions between $y$ parameters.

emitter $S$ parameters to common-emitter $y$ parameters, then convert the common-emitter $y$ parameters to common-base $y$ parameters, and then convert the common-base $y$ parameters to common-base $S$ parameters.

The frequency characteristics of a network can be represented as a continuous impedance or reflection coefficient plot in the Smith chart (see Sections 2.2 and 2.3). For example, the series $RC$ network shown in Fig. 1.9.1a has the impedance or reflection coefficient plot shown in Fig. 1.9.1b. As the frequency increases, the capacitive reactance decreases, and the impedance plot moves clockwise along a constant-resistance circle.

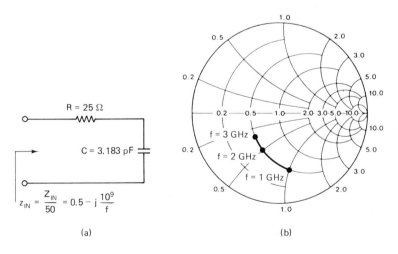

**Figure 1.9.1** Frequency response of a series $RC$ network.

Sec. 1.9  Scattering Parameters of Transistors

A typical plot of $S_{11}$ for a transistor in the common-emitter configuration is shown in Fig. 1.9.2. The plot of $S_{11}$ is given for the transistor in chip form and in packaged form. The bias conditions are also shown. It is observed that $S_{11}$ for this transistor in chip form follows a constant-resistance circle, with a capacitive reactance at the lower frequencies and an inductive reactance at the higher frequencies. The equivalent circuit for this transistor in chip form which exhibits the behavior of $S_{11}$ is shown in Fig. 1.9.3a. The resistance $R$ represents the base-to-emitter resistance plus any contact resistance. The capacitance $C$ is due to the junction capacitance from base to emitter. The inductance $L$ is due to the reflection properties of a transistor where the emitter resistance, when $h_{fe}(\omega)$ is complex, produces an inductive reactance across the base-to-emitter terminals.

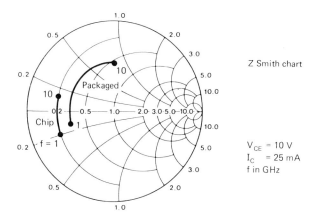

**Figure 1.9.2**  $S_{11}$ of a common-emitter transistor in chip and packaged form.

The equivalent circuit for the transistor in packaged form is shown in Fig. 1.9.3b. In this case, the package inductance ($L_{pkg}$) and the package capacitance ($C_{pkg}$) contribute to the reflection coefficient variations at the higher frequencies.

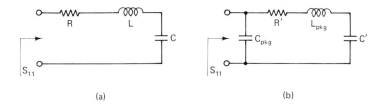

**Figure 1.9.3**  (a) Input equivalent circuit for a transistor in chip form; (b) input equivalent circuit for a packaged transistor.

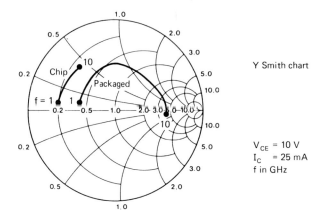

**Figure 1.9.4** $S_{22}$ of a common-emitter transistor in chip and packaged form.

A typical $S_{22}$ plot for a chip and packaged transistor in the common-emitter configuration is shown in Fig. 1.9.4. For this transistor the chip characteristic follows a constant conductance curve (i.e., a shunt $RC$ equivalent network).

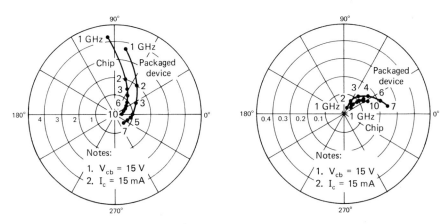

**Figure 1.9.5** $S_{21}$ of a common-emitter transistor in chip and packaged form. (From Ref. [1.1]; courtesy of Hewlett-Packard.)

**Figure 1.9.6** $S_{12}$ of a common-emitter transistor in chip and packaged form. (From Ref. [1.1]; courtesy of Hewlett-Packard.)

The forward and reverse transmission coefficients $S_{21}$ and $S_{12}$ are usually given in a polar plot as shown in Figs. 1.9.5 and 1.9.6.

The parameter $|S_{21}|$ is constant for frequencies below the beta cutoff frequency (i.e., $f_\beta$) and then decays at 6 dB/octave. The transducer cutoff frequency ($f_s$) is the frequency where $|S_{21}|$ is equal to 1. The parameter $|S_{12}|$

## Sec. 1.9 Scattering Parameters of Transistors

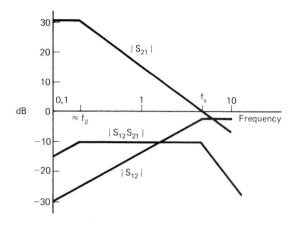

**Figure 1.9.7** Frequency behavior of $|S_{21}|$, $|S_{12}|$, and $|S_{12}S_{21}|$. (From Ref. [1.1]; courtesy of Hewlett-Packard.)

increases at approximately 6 dB/octave, levels off around $f_s$, and decays at higher frequencies. A typical Bode plot of $|S_{21}|$, $|S_{12}|$, and the product $|S_{12}S_{21}|$ is shown in Fig. 1.9.7.

The common-emitter $S$ parameters of a transistor are shown in Fig. 1.9.8. This figure illustrates some of the information typically provided by manufacturers.

A transistor can be considered to be a three-port device as shown in Fig. 1.9.9. In this case the scattering matrix, also called the *indefinite scattering matrix*, is

$$\begin{bmatrix} b_1 \\ b_2 \\ b_3 \end{bmatrix} = \begin{bmatrix} S_{11} & S_{12} & S_{13} \\ S_{21} & S_{22} & S_{23} \\ S_{31} & S_{32} & S_{33} \end{bmatrix} \begin{bmatrix} a_1 \\ a_2 \\ a_3 \end{bmatrix} \qquad (1.9.1)$$

The name "indefinite scattering matrix" is used because no definite choice is made to ground a particular port. The meaning of $S_{11}$ in (1.9.1) is

$$S_{11} = \left. \frac{b_1}{a_1} \right|_{a_2 = 0, \, a_3 = 0} \qquad (1.9.2)$$

That is, to measure $S_{11}$ reference resistances of 50 $\Omega$ are used at ports 2 and 3. In a two-port common-emitter configuration, $S_{11}$ is measured with the emitter grounded. Therefore, the value of $S_{11}$ in (1.9.2) will be different from the value of $S_{11}$ in a two-port common-emitter configuration. Similarly, the parameter $S_{12}$, $S_{21}$, and $S_{22}$ in (1.9.1) will be different from the $S_{12}$, $S_{21}$, and $S_{22}$ in a two-port common-emitter configuration.

## MRF962 COMMON-EMITTER S-PARAMETERS

INPUT/OUTPUT REFLECTION
COEFFICIENTS versus FREQUENCY
($V_{CE}$ = 10 V, $I_C$ = 50 mA)

FORWARD/REVERSE TRANSMISSION
COEFFICIENTS versus FREQUENCY
($V_{CE}$ = 10 V, $I_C$ = 50 mA)

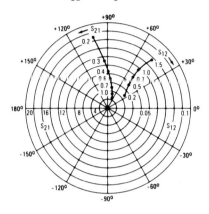

| $V_{CE}$ (Volts) | $I_C$ (mA) | f (MHz) | $S_{11}$ | | $S_{21}$ | | $S_{12}$ | | $S_{22}$ | |
|---|---|---|---|---|---|---|---|---|---|---|
| | | | $|S_{11}|$ | $\angle \phi$ | $|S_{21}|$ | $\angle \phi$ | $|S_{12}|$ | $\angle \phi$ | $|S_{22}|$ | $\angle \phi$ |
| 5.0 | 10 | 100  | 0.70 | -102 | 17.42 | 128 | 0.044 | 43 | 0.65 | -57 |
|     |    | 300  | 0.75 | -156 | 7.11  | 98  | 0.058 | 24 | 0.32 | -97 |
|     |    | 500  | 0.78 | -170 | 4.36  | 86  | 0.064 | 25 | 0.26 | -110 |
|     |    | 700  | 0.78 | -176 | 3.16  | 77  | 0.071 | 26 | 0.23 | -117 |
|     |    | 1000 | 0.78 | 176  | 2.26  | 67  | 0.078 | 27 | 0.24 | -126 |
|     |    | 1500 | 0.79 | 167  | 1.51  | 54  | 0.092 | 29 | 0.31 | -133 |
|     | 25 | 100  | 0.69 | -131 | 24.24 | 118 | 0.029 | 38 | 0.56 | -87 |
|     |    | 300  | 0.77 | -167 | 8.76  | 95  | 0.039 | 32 | 0.35 | -137 |
|     |    | 500  | 0.79 | -176 | 5.26  | 85  | 0.046 | 36 | 0.32 | -150 |
|     |    | 700  | 0.80 | 178  | 3.82  | 78  | 0.055 | 40 | 0.31 | -158 |
|     |    | 1000 | 0.79 | 173  | 2.72  | 70  | 0.067 | 42 | 0.32 | -164 |
|     |    | 1500 | 0.81 | 164  | 1.82  | 59  | 0.086 | 42 | 0.34 | -167 |
|     | 50 | 100  | 0.71 | -147 | 27.72 | 113 | 0.021 | 37 | 0.53 | -107 |
|     |    | 300  | 0.78 | -173 | 9.59  | 94  | 0.030 | 40 | 0.41 | -152 |
|     |    | 500  | 0.81 | 179  | 5.72  | 85  | 0.038 | 46 | 0.39 | -163 |
|     |    | 700  | 0.81 | 176  | 4.09  | 78  | 0.048 | 50 | 0.38 | -169 |
|     |    | 1000 | 0.81 | 171  | 2.89  | 71  | 0.061 | 51 | 0.38 | -175 |
|     |    | 1500 | 0.82 | 163  | 1.96  | 62  | 0.082 | 49 | 0.40 | -177 |
| 10  | 10 | 100  | 0.71 | -92  | 18.77 | 131 | 0.037 | 47 | 0.70 | -44 |
|     |    | 300  | 0.74 | -150 | 8.09  | 100 | 0.051 | 28 | 0.34 | -69 |
|     |    | 500  | 0.75 | -166 | 5.01  | 87  | 0.056 | 28 | 0.27 | -75 |
|     |    | 700  | 0.76 | -174 | 3.62  | 78  | 0.064 | 28 | 0.24 | -79 |
|     |    | 1000 | 0.76 | 179  | 2.58  | 69  | 0.071 | 30 | 0.24 | -88 |
|     |    | 1500 | 0.77 | 168  | 1.72  | 55  | 0.085 | 31 | 0.31 | -104 |
|     | 25 | 100  | 0.67 | -120 | 27.10 | 122 | 0.027 | 42 | 0.57 | -68 |
|     |    | 300  | 0.73 | -163 | 10.27 | 97  | 0.035 | 36 | 0.27 | -110 |
|     |    | 500  | 0.76 | -174 | 6.21  | 86  | 0.043 | 39 | 0.22 | -124 |
|     |    | 700  | 0.77 | -179 | 4.48  | 78  | 0.051 | 41 | 0.20 | -132 |
|     |    | 1000 | 0.77 | 175  | 3.19  | 71  | 0.062 | 43 | 0.20 | -139 |
|     |    | 1500 | 0.78 | 166  | 2.13  | 59  | 0.080 | 42 | 0.25 | -142 |
|     | 50 | 100  | 0.68 | -137 | 31.53 | 116 | 0.020 | 37 | 0.49 | -85 |
|     |    | 300  | 0.74 | -169 | 11.17 | 95  | 0.028 | 40 | 0.27 | -131 |
|     |    | 500  | 0.77 | -177 | 6.69  | 85  | 0.037 | 46 | 0.24 | -144 |
|     |    | 700  | 0.77 | 178  | 4.82  | 78  | 0.047 | 48 | 0.23 | -152 |
|     |    | 1000 | 0.77 | 173  | 3.42  | 71  | 0.059 | 50 | 0.23 | -158 |
|     |    | 1500 | 0.79 | 165  | 2.30  | 61  | 0.078 | 47 | 0.27 | -159 |

**Figure 1.9.8** S-parameter data for the Motorola MRF962 transistor. (From Motorola RF Data Manual; reproduced with permission of Motorola, Inc.)

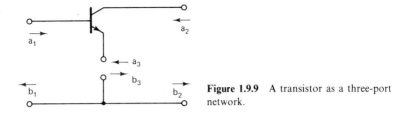

**Figure 1.9.9** A transistor as a three-port network.

## 1.10 CHARACTERISTICS OF MICROWAVE TRANSISTORS

Most microwave bipolar junction transistors (BJTs) are planar in form and made from silicon in the *npn* type. Their dimensions are very small to permit operation at microwave frequencies. For example, the base thickness is of the order of a tenth of a micron.

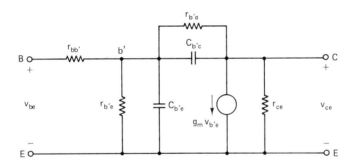

**Figure 1.10.1** Hybrid-$\pi$ model of a common-emitter BJT.

The equivalent hybrid-$\pi$ model of the intrinsic BJT is shown in Fig. 1.10.1. In the microwave range, the reactance of $C_{b'c}$ is very small in comparison to the resistance of $r_{b'c}$, and the resistor $r_{ce}$ is very large. Also, the reactance of $C_{b'e}$ is usually smaller than the resistance $r_{b'e}$. Therefore, the simplified model shown in Fig. 1.10.2 follows.

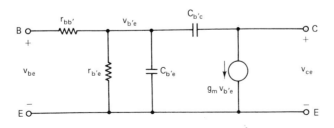

**Figure 1.10.2** Simplified hybrid-$\pi$ model of a microwave BJT in the common-emitter configuration.

In the actual (or extrinsic) microwave transistor, additional parasitic resistances, inductances, and capacitances appear (as well as other distributed elements) and must be included in the model. One such model is shown in Fig. 1.10.3. The meaning of the parasitic elements $R_e$, $L_b$, $L_e$, $L_c$, $C_{be}$, and $C_{ce}$ is self-explanatory.

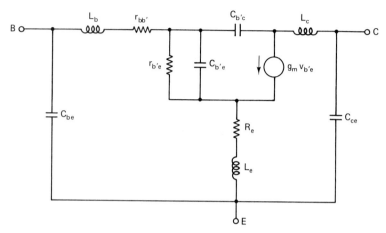

**Figure 1.10.3** A microwave BJT common-emitter model that includes parasitics.

There are two figures of merit that are commonly used by manufacturers of microwave BJTs to describe the transistor performance. These are:

1. $f_T$: the gain–bandwidth frequency. It is the frequency where the short-circuit gain $|h_{fe}(\omega)|$ approximates unity.
2. $f_{max}$: the maximum frequency of oscillation. It is the frequency where the maximum available power gain of the transistor (called $G_{A,max}$) is equal to 1.

$G_{A,max}$ and $f_{max}$ can be measured by conjugately matching the source impedance to the transistor input impedance, and the load to the transistor output impedance. Of course, the transistor must be unconditionally stable (i.e., no oscillations). $G_{A,max}$ is higher than the transducer gain $|S_{21}|^2$ because of the matching conditions. These concepts are discussed in detail in Chapter 3.

The frequency dependence of $h_{fe}(\omega)$ is given by

$$h_{fe}(\omega) = \frac{h_{fe}}{1 + jf/f_\beta}$$

where $h_{fe}$ is the low-frequency short-circuit current gain and $f_\beta$ is the beta

## Sec. 1.10 Characteristics of Microwave Transistors

cutoff frequency, namely

$$f_\beta = \frac{1}{2\pi r_{b'e}(C_{b'e} + C_{b'c})} \approx \frac{1}{2\pi r_{b'e} C_{b'e}}$$

The frequencies $f_T$ and $f_{max}$ for the intrinsic BJT model, shown in Fig. 1.10.2, are given by

$$f_T \approx \frac{g_m}{2\pi C_{b'e}} \quad (1.10.1)$$

and

$$f_{max} = \sqrt{\frac{f_T}{8\pi r_{b'e} C_{b'c}}} \quad (1.10.2)$$

Also, $f_\beta$ and $f_T$ are related by

$$f_\beta = \frac{f_T}{h_{fe}} \quad (1.10.3)$$

The frequency $f_T$ can also be expressed in terms of the total signal time delay from emitter to collector as [1.2]

$$f_T = \frac{1}{2\pi \tau_{ec}}$$

where $\tau_{ec}$ is the emitter-to-collector time delay, namely

$$\tau_{ec} \approx \tau_b + \tau_c$$

The parameter $\tau_b$ represents the base delay time and $\tau_c$ the base-to-collector depletion-layer delay time.

Figure 1.10.4 illustrates the meaning of $f_T$, $f_s$, and $f_{max}$. Observe the gain rolloff at the rate of 6 dB/octave.

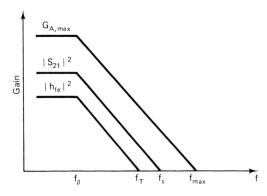

**Figure 1.10.4** Frequency characteristics of $G_{A,max}$, $|S_{21}|^2$ and $|h_{fe}|^2$.

Two sources of noise in a microwave BJT are thermal noise and shot noise. *Thermal noise* is caused by the thermal agitation of the carriers in the ohmic resistance of the emitter, base, and collector. *Shot noise* is a current dependent effect caused by fluctuations in the electron and hole currents due to bias conditions.

The gallium arsenide (GaAs) field-effect transistors (FETs) are commonly made in the metal semiconductor field-effect transistor structure (MESFET). That is, the gate terminal is constructed using a Schottky barrier gate. However, the other FET structures, such as the junction field-effect transistor (JFET), the metal-oxide semiconductor field-effect transistor (MOSFET), and the insulated-gate field-effect transistor (IGFET), have also been used.

The microwave FETs are made with GaAs because the electron mobility is greater than that of silicon. The high electron mobility results in excellent frequency response and noise performance, especially above 4 GHz.

The high-frequency model of the intrinsic GaAs FET in a common-source configuration is shown in Fig. 1.10.5. The capacitor $C_i$ represents the gate-to-source capacitance and the resistor $r_i$ is the small gate-to-source channel resistance. An extrinsic GaAs FET high-frequency model which includes parasitic elements is shown in Fig. 1.10.6. The meaning of the parasitic elements $R_g$, $R_s$, $R_d$, $L_g$, $L_s$, and $L_d$ is self-explanatory.

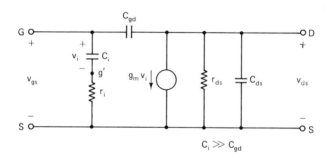

**Figure 1.10.5** GaAs FET high-frequency model for the common-source configuration.

When the GaAs FET is unconditionally stable in the frequency range of interest, the feedback capacitance is very small and can be neglected. That is, there is no reverse transmission from the output to the input port of the transistor and the transistor becomes unilateral (i.e., $S_{12} = 0$). The simplified unilateral high-frequency model for the intrinsic GaAs FET is shown in Fig. 1.10.7.

The short length of the gate determines the frequency response of the GaAs FET. For the model shown in Fig. 1.10.7, the frequencies $f_T$ and $f_{max}$ are given by

$$f_T = \frac{g_m}{2\pi C_i} \qquad (1.10.4)$$

## Sec. 1.10 Characteristics of Microwave Transistors

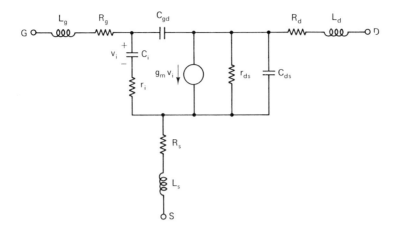

**Figure 1.10.6** A microwave GaAs FET common-source model that includes parasitic elements.

and

$$f_{max} = \frac{f_T}{2}\sqrt{\frac{r_{ds}}{r_i}} \qquad (1.10.5)$$

Since $f_T$ is limited by the electron transit time $(\tau_c)$ through the channel, $f_T$ can be expressed in the form [1.2]

$$f_T = \frac{1}{2\pi\tau_c}$$

where

$$\tau_c = \frac{L}{v_s}$$

Here $L$ is the gate length and $v_s$ is the electron saturation drift velocity.

Another expression for $f_{max}$, found experimentally, is [1.2]

$$f_{max} = \frac{33 \times 10^3}{L}$$

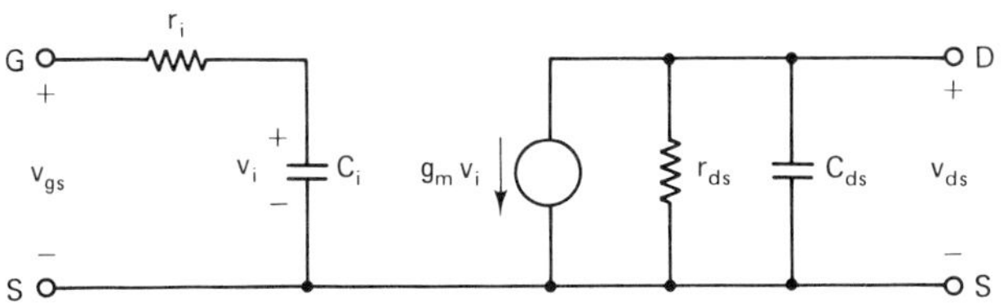

**Figure 1.10.7** Simplified unilateral high-frequency model for a common-source GaAs FET.

The intrinsic noise sources in a GaAs FET are the thermal-generated channel noise and the induced noise at the gate. The induced noise at the gate is produced by the channel noise voltage. The extrinsic noise sources (see Fig. 1.10.6) are associated with the resistances $R_g$ and $R_s$, and the gate bonding pad resistance. In addition, the GaAs FET exhibits intervalley scattering noise produced by scattering of electrons between energy bands.

## PROBLEMS

**1.1.** A section $\Delta x$ of a uniform transmission line is shown in Fig. P1.1.
   (a) Using Kirchhoff's voltage and current laws, show that as $\Delta x \to 0$

$$-\frac{\partial v(x,t)}{\partial x} = Ri(x,t) + L\frac{\partial i(x,t)}{\partial t}$$

and

$$-\frac{\partial i(x,t)}{\partial x} = Gv(x,t) + C\frac{\partial v(x,t)}{\partial t}$$

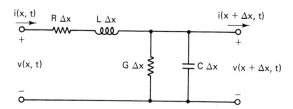

Figure P1.1

(b) Show that $v(x,t)$ and $i(x,t)$ satisfy the equations

$$\frac{\partial^2 v(x,t)}{\partial x^2} = LC\frac{\partial^2 v(x,t)}{\partial t^2} + (RC + LG)\frac{\partial v(x,t)}{\partial t} + RGv(x,t)$$

and

$$\frac{\partial^2 i(x,t)}{\partial x^2} = LC\frac{\partial^2 i(x,t)}{\partial t^2} + (RC + LG)\frac{\partial i(x,t)}{\partial t} + RGi(x,t)$$

   (c) Assuming a sinusoidal excitation, show that the phasors $V(x)$ and $I(x)$ satisfy (1.3.1) and (1.3.2), respectively.
   (d) Verify the general solution for $V(x)$ and $I(x)$ in (1.3.3) and (1.3.4).
   (e) Show that in a lossless transmission line the voltage $V(x)$ satisfies

$$\frac{d^2 V(x)}{dx^2} + \beta^2 V(x) = 0$$

where $\beta = \omega\sqrt{LC}$. Write the general solution for $V(x)$.

**1.2.** Verify the equations for $Z_{IN}(d)$ in (1.3.9) and (1.3.10).

**1.3.** Verify the $S$- and $T$-parameter conversions given in (1.4.7) and (1.4.8).

**1.4.** (a) Find the *ABCD* matrix of the series impedance $Z$ and shunt admittance $Y$ of Example 1.6.1.
(b) Use Fig. 1.8.1 to convert the *ABCD* parameters obtained in part (a) to $S$ parameters. Compare the answers with the results in Example 1.6.1.

**1.5.** Find the scattering matrix and the chain scattering matrix of a transmission line of length $l$ and characteristic impedance $Z_o$.

**1.6.** Find the scattering matrix and the chain scattering matrix of
(a) A short-circuited shunt stub of length $l$ and characteristic impedance $Z_o$.
(b) An open-circuited shunt stub of length $l$ and characteristic impedance $Z_o$.

**1.7.** Find the $S$ parameters of the 1-to-$n$ turns ratio transformer shown in Fig. P1.7 at ports 1–2 and 1′–2′.

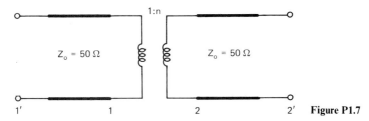

**Figure P1.7**

**1.8.** Show that the overall $S_{21}$ parameter of two cascade two-port networks with scattering matrices $[S_A]$ and $[S_B]$, respectively, is given by

$$S_{21} = \frac{S_{21,A} S_{21,B}}{1 - S_{22,A} S_{11,B}}$$

**1.9.** In the two-port network shown in Fig. P1.9:
(a) Find $Z_{IN}(0)$.
(b) Find $a_1(0)$, $b_1(0)$, $a_1(\lambda/8)$, $b_1(\lambda/8)$, and $a_2(0)$.
(c) Evaluate $V_1(0)$, $V_1(\lambda/8)$, $I_1(0)$, and $I_1(\lambda/8)$.
(d) Evaluate the average input power at $x_1 = 0$ and the average input power at $x_1 = \lambda/8$.
(e) Show that $P_1(0) = P_1(\lambda/8)$.
(f) Evaluate $S_{11}(0)$ and $S_{11}(\lambda/8)$.
(g) Find the electrical length and the length in centimeters of the $\lambda/8$ transmission line at $f = 1$ GHz.

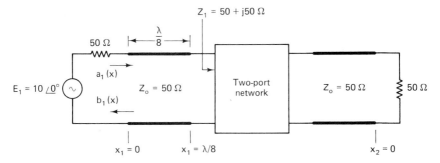

**Figure P1.9**

**1.10.** (a) Find the value of the source impedance that results in maximum power delivered to the load in Fig. P1.10. Evaluate the maximum power delivered to the load.
(b) Using the value of $Z_s$ from part (a), find the Thévenin's equivalent circuit at the load end and evaluate the power delivered to the load.

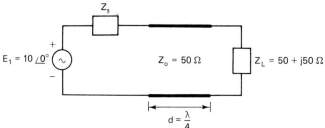

**Figure P1.10**

**1.11.** Show that in the $n$-port network of Fig. 1.7.1, the transducer power gain from port $i$ to port $k$, $|S_{ki}|^2$, is given by

$$|S_{ki}|^2 = \frac{\text{Re}(Z_{o,k})|I_k|^2}{|E_i|^2/4\text{Re}(Z_{o,i})} = \frac{P_k}{(P_{\text{AVS}})_i}$$

where $(P_{\text{AVS}})_i$ is the maximum available power at port $i$ and $P_k$ is the power delivered to $Z_{o,k}$.

**1.12.** In Fig. 1.8.1, verify the conversions between
(a) $z$ and $y$ parameters
(b) $z$ and $ABCD$ parameters

**1.13.** (a) Show that

$$[S] = -([y] + [Y_o])^{-1}([y] - [Y_o])$$

and

$$[y] = [Y_o]([1] - [S])([1] + [S])^{-1}$$

where

$$[Y_o] = \begin{bmatrix} Y_o & 0 \\ 0 & Y_o \end{bmatrix}$$

(b) Verify the conversion between $S$ and $y$ parameters in Fig. 1.8.1.

**1.14.** In the network shown in Fig. P1.14, the $S$ parameters of the BJT and the value of $L$ are known. Explain how the overall $S$ parameters of the two-port can be calculated.

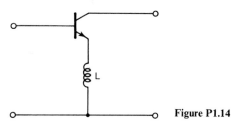

**Figure P1.14**

**1.15.** The common-emitter S-parameters of a GaAs FET at $f = 10$ GHz are

$$S_{11} = 0.73 \underline{|-128°}$$
$$S_{21} = 1.73 \underline{|73°}$$
$$S_{12} = 0.045 \underline{|114°}$$
$$S_{22} = 0.75 \underline{|-52°}$$

Determine the common-base and common-collector S-parameters.

**1.16.** Find equivalent circuits that exhibit the $S_{22}$ characteristics (chip and packaged form) shown in Fig. 1.9.4.

**1.17.** (a) Derive the equations for $f_T$, $f_{max}$, and $f_\beta$, for a BJT, given in (1.10.1), (1.10.2), and (1.10.3).

(b) Derive the equations for $f_T$ and $f_{max}$, for a GaAs FET, given in (1.10.4) and (1.10.5).

**1.18.** Show that for equal reference resistance $(Z_{o1} = Z_{o2})$, the scattering matrix in (1.4.5) can be written in the form

$$V_1^- = S_{11}V_1^+ + S_{12}V_2^+$$
$$V_2^- = S_{21}V_1^+ + S_{22}V_2^+$$

and

$$I_1^- = S_{11}I_1^+ + S_{12}I_2^+$$
$$I_2^- = S_{21}I_1^+ + S_{22}I_2^+$$

**1.19.** Show that in the indefinite scattering matrix given in (1.9.1), the sum of the coefficient of any row is equal to 1 and the sum of the coefficient of any column is equal to 1.

*Hint*: Since (1.9.1) is valid for any values of $a_1$, $a_2$, and $a_3$, consider the case where $a_2 = a_3 = 0$ as shown in Fig. P1.19a. Then $b_1 = S_{11}a_1$, $b_2 = S_{21}a_1$, and $b_3 = S_{31}a_1$, and at $P$ we can write

$$I_1^+ = I_1^- + I_2^- + I_3^-$$

Therefore, it follows that

$$S_{11} + S_{21} + S_{31} = 1$$

The circuit shown in Fig. P1.19b can be used to show that the sum of the coefficients in any row is equal to 1.

**1.20.** Show that the generalized scattering parameters for a two-port network in terms of arbitrary source $(Z_s)$ and load $(Z_L)$ impedances are given by

$$S'_{11} = \frac{A_1^* (1 - \Gamma_L S_{22})(S_{11} - \Gamma_s^*) + \Gamma_L S_{12} S_{21}}{A_1 \quad D}$$

$$S'_{12} = \frac{A_2^* S_{12}[1 - |\Gamma_s|^2]}{A_2 \quad D}$$

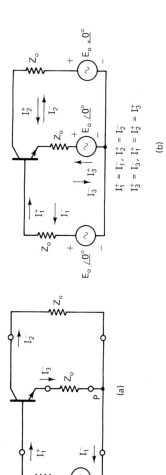

**Figure P1.19**

$$S'_{21} = \frac{A_1^*}{A_1} \frac{S_{21}[1 - |\Gamma_L|^2]}{D}$$

$$S'_{22} = \frac{A_2^*}{A_2} \frac{(1 - \Gamma_s S_{22})(S_{22} - \Gamma_L^*) + \Gamma_s S_{12} S_{22}}{D}$$

where

$$D = [(1 - \Gamma_s S_{11})(1 - \Gamma_L S_{22}) - \Gamma_s \Gamma_L S_{12} S_{21}]$$

$$A_1 = \frac{1 - \Gamma_s^*}{|1 - \Gamma_s|} \sqrt{1 - |\Gamma_s|^2}, \quad A_2 = \frac{(1 - \Gamma_L^*)}{|1 - \Gamma_L|} \sqrt{1 - |\Gamma_L|^2}$$

$$\Gamma_s = \frac{Z_s - Z_o}{Z_s + Z_o}, \quad \Gamma_L = \frac{Z_L - Z_o}{Z_L + Z_o}$$

The S parameters are normalized to a real $Z_o$.

*Note:* This is a difficult problem that involves some matrix manipulations to obtain $[S']$ as a function of $[S]$.

## REFERENCES

[1.1] "S Parameter Design," Hewlett-Packard Application Note 154, April 1972.

[1.2] D. V. Morgan and M. J. Howes, editors, *Microwave Solid State Devices and Applications*, Peter Peregrinus Ltd., New York, 1980.

# 2
# MATCHING NETWORKS AND SIGNAL FLOW GRAPHS

## 2.1 INTRODUCTION

The analysis of transmission-line problems and of matching circuits at microwave frequencies can be cumbersome in analytical form. The Smith chart provides a very useful graphical aid to the analysis of these problems. The Smith chart is basically a plot of all passive impedances in a reflection coefficient chart of unit radius. The reading accuracy from the Smith chart is sufficient for most practical microwave transistor amplifier design problems.

Matching circuits that provide optimum performance in a microwave amplifier can be easily and quickly designed using the normalized impedance and admittance Smith chart. The Smith chart is also used to present the frequency dependence of scattering parameters and other amplifier characteristics.

The characteristics of microstrip transmission lines are presented in this chapter. The mode of propagation in a microstrip line is assumed to be quasi-transverse electromagnetic. Although radiation losses in a microstrip line can be severe, the use of a thin material, having a high dielectric constant, between the top strip conductor and the ground plane of a microstrip line reduces the radiation losses to a minimum.

Microstrip lines find extensive use as passive circuit elements and as a medium in which the complete microwave amplifier can be built. The interconnection features of the microstrip line are unsurpassed. Transistors in chip or packaged form can be easily attached to the strip conductors of the microstrip line. Some practical circuit construction techniques using microstrips are presented.

Sec. 2.2   The Smith Chart

The description of two-port networks in terms of $S$ parameters permits the use of signal flow graph in the analysis of microwave transistor amplifiers.

## 2.2 THE SMITH CHART

The Smith chart is the representation in the reflection coefficient plane, called the $\Gamma$ *plane*, of the relation

$$\Gamma = \frac{Z - Z_o}{Z + Z_o} \quad (2.2.1)$$

for all values of $Z$, such that Re $[Z] \geq 0$. $Z_o$ is the characteristic impedance of the transmission line or a reference impedance value. Defining the normalized impedance $z$ as $z = Z/Z_o$, we can write (2.2.1) in the form

$$\Gamma = \frac{z - 1}{z + 1} \quad (2.2.2)$$

Figure 2.2.1 illustrates the properties of the transformation (2.2.2) for some values of $z$. For example, if $Z = 50 \, \Omega$ and $Z_o = 50 \, \Omega$, then $z = 1$ and $\Gamma = 0$. That is, the point $z = 1$ in the normalized $z$ plane maps into the origin of the $\Gamma$ plane. Next, we consider the mapping of normalized impedances having constant real and imaginary parts. For example, for $z = 1 + jx$ the corresponding values of $\Gamma$ will be shown to lie in a circle of radius $1/2$, centered at $U = 1/2$, $V = 0$. Also, for $z = r + j1$ ($r \geq 0$) it follows that the corresponding values of $\Gamma$ lie in a circle of radius 1, centered at $U = 1$, $V = 1$. All passive impedances, that is, impedances having $r \geq 0$ map inside the unit circle (i.e. $|\Gamma| \leq 1$) in the $\Gamma$ plane. Observe that the imaginary axis (i.e., $r = 0$) maps into the unit circle, $|\Gamma| = 1$.

The transformation (2.2.2) can be analyzed in general as follows. Let

$$\Gamma = U + jV = \frac{(r - 1) + jx}{(r + 1) + jx}$$

Then rationalize and separate the real and imaginary parts to obtain

$$U = \frac{r^2 - 1 + x^2}{(r + 1)^2 + x^2} \quad (2.2.3)$$

and

$$V = \frac{2x}{(r + 1)^2 + x^2} \quad (2.2.4)$$

Eliminating $x$ from (2.2.3) and (2.2.4) results in

$$\left(U - \frac{r}{r + 1}\right)^2 + V^2 = \left(\frac{1}{r + 1}\right)^2$$

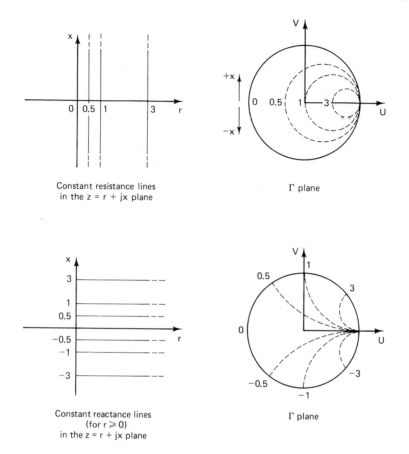

**Figure 2.2.1** Development of the Smith chart.

which is the equation of a family of circles centered at $U = r/(r+1)$, $V = 0$ with radii $1/(r+1)$. The circles for $r = 0, 0.5, 1$, and $3$ are shown in Fig. 2.2.1. Eliminating $r$ from (2.2.3) and (2.2.4) results in

$$(U-1)^2 + \left(V - \frac{1}{x}\right)^2 = \left(\frac{1}{x}\right)^2$$

which is the equation of a family of circles centered at $U = 1$, $V = 1/x$, with radii $1/x$. The circles for $x = -3, -1, -0.5, 0, 1, 0.5$, and $3$ (with $r \geq 0$) are shown in Fig. 2.2.1.

There is a one-to-one correspondence between points in the $z$ plane and points in the $\Gamma$ plane. The plot of the constant-resistance and constant-reactance circles in a graph is known as the *Smith chart*. The Smith chart is shown in Fig. 2.2.2. Observe that the upper half of the chart represents normalized impedances having a positive reactance (i.e., $+jx$), and the lower half represents negative reactances (i.e., $-jx$). The distance around the chart is $\lambda/2$.

## Sec. 2.2 The Smith Chart

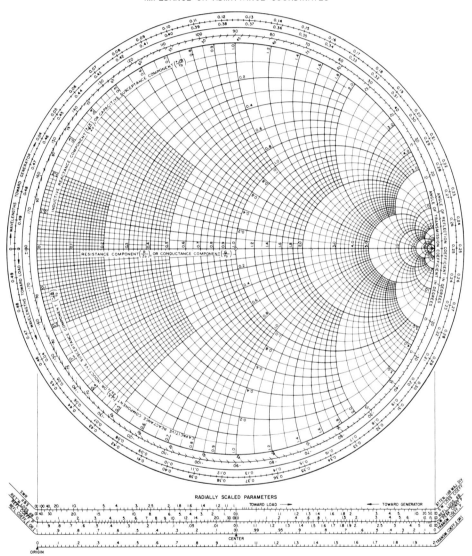

**Figure 2.2.2** The Smith chart. (Reproduced with permission of Kay Electric Co., Pine Brook, N.J.)

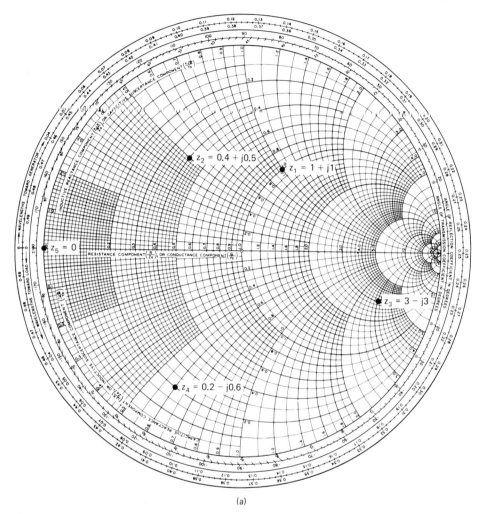

(a)

**Figure 2.2.3** (a) Values of $z$ in the $Z$ Smith chart; (b) values of $y$ in the $Y$ Smith chart.

The Smith chart can also be used as an admittance chart. The appropriate transformation in this case is

$$\Gamma = \frac{y-1}{y+1}$$

where the normalized admittance is $y = Y/Y_o$. $Y_o$ is the characteristic admittance of the transmission line or a reference admittance value.

In the admittance chart, since $y = g + jb$, the previous constant-resistance ($r$) circles become constant-conductance ($g$) circles and the constant-reactance ($x$) circles become constant-susceptance ($b$) circles. Observe that the upper half of the chart represents normalized admittances having a positive

Sec. 2.2  The Smith Chart

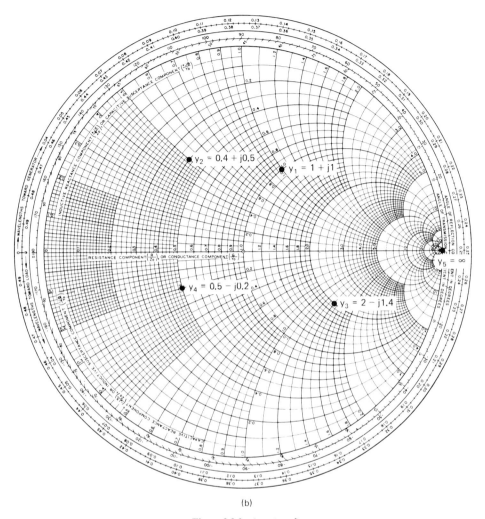

(b)

**Figure 2.2.3** (*continued*)

susceptance (i.e., $+jb$) and the lower half represents negative susceptances (i.e., $-jb$).

When needed for clarity, we will call a Smith chart used as an impedance chart a "Z Smith chart," and a Smith chart used as an admittance chart a "Y Smith chart."

**Example 2.2.1**

Locate in the Smith chart the following normalized impedances and admittances:

$$z_1 = 1 + j1, \quad z_2 = 0.4 + j0.5, \quad z_3 = 3 - j3, \quad z_4 = 0.2 - j0.6, \quad z_5 = 0$$

$$y_1 = 1 + j1, \quad y_2 = 0.4 + j0.5, \quad y_3 = 2 - j1.4, \quad y_4 = 0.5 - j0.2, \quad y_5 = \infty$$

**Solution.** The values of $z$'s and $y$'s are shown in Fig. 2.2.3. The Smith chart in Fig. 2.2.3a is obviously used as a Z Smith chart, and that in Fig. 2.2.3b as a Y Smith chart.

Impedances having a negative real part will have a reflection coefficient whose magnitude is greater than 1. These impedances, therefore, map outside the Smith chart. Figure 2.2.4 shows a chart (known as the *compressed Smith chart*) that includes the Smith chart (i.e., $|\Gamma| \le 1$) plus a portion of the negative impedance region.

An alternative way of handling negative resistances (i.e., $|\Gamma| > 1$) is to plot in the Smith chart $1/\Gamma^*$ and take the values of the resistance circles as being negative and the reactance circles as labeled.

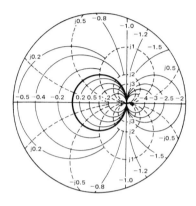

**Figure 2.2.4** The compressed Smith chart. (From Ref. [1.1]; courtesy of Hewlett-Packard.)

**Example 2.2.2**

Find the impedance whose reflection coefficient is $2.236e^{j26.56°}$.

**Solution.** If we plot in the Smith chart shown in Fig. 2.2.5, the quantity

$$\frac{1}{\Gamma^*} = 0.447e^{j26.56°}$$

the resulting $z$ is $-2 + j1$. Of course, from (2.2.2),

$$\Gamma = \frac{-2 + j1 - 1}{-2 + j1 + 1} = 2.236e^{j26.56°}$$

The use of the Smith chart in a transmission-line calculation follows from (1.3.5), (1.3.7), and (1.3.8). For a lossless transmission line, we can conveniently write (1.3.5), (1.3.7), and (1.3.8) in the form

$$\Gamma_0 = \frac{z-1}{z+1} \qquad (2.2.5)$$

$$\Gamma_{IN}(d) = \Gamma_0 e^{-j2\beta d} \qquad (2.2.6)$$

$$z_{IN}(d) = \frac{1 + \Gamma_{IN}(d)}{1 - \Gamma_{IN}(d)} \qquad (2.2.7)$$

Sec. 2.2  The Smith Chart   49

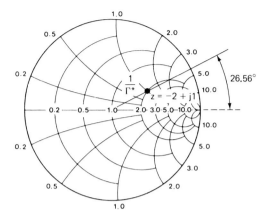

**Figure 2.2.5** Negative resistances in the Smith chart.

A typical transmission-line input impedance calculation involves the following steps:

1. Locate $\Gamma_0$ in the Z Smith chart for a given $z = Z_L/Z_o$ [i.e., (2.2.5)].
2. Rotate $\Gamma_0$ by $-2\beta d$ to obtain $\Gamma_{IN}(d)$ [i.e., (2.2.6)]. Observe that the rotation is along a vector of constant magnitude, namely $|\Gamma_0| = |\Gamma_{IN}(d)|$.
3. Read the value of the normalized $z_{IN}(d)$ associated with $\Gamma_{IN}(d)$ [i.e., (2.2.7)].

**Example 2.2.3**

Find the input impedance, the reflection coefficient, and the VSWR in a transmission line having an electrical length of 45°, characteristic impedance of 50 Ω, and terminated in a load $Z_L = 50 + j50$ Ω.

**Solution.** The transmission line is shown in Fig. 2.2.6a, where $z_L = Z_L/Z_o = 1 + j1$ and $\beta d = 2\pi d/\lambda = \pi/4$ or $d = \lambda/8 = 0.125\lambda$. In Fig. 2.2.6b the point $z_L = 1 + j1$ is located and the vector representing $\Gamma_0$ drawn. To find $Z_{IN}$, we rotate along a constant $\Gamma$ radius a distance of $-90°$ (i.e., $-2\beta d$). Observe that the Smith chart already has a wavelength scale. Therefore, if we rotate toward the generator a distance $d = 0.125\lambda$ along the wavelength scale, it is equivalent to a rotation of $-90°$. The procedure is illustrated in Fig. 2.2.6b. That is, at $z = 1 + j1$, we read the value of $0.162\lambda$ from the wavelength scale; next, we add $0.125\lambda$ and rotate along a constant $|\Gamma_0|$ radius until we reach the value of $0.162\lambda + 0.125\lambda = 0.287\lambda$. At $0.287\lambda$ the input impedance is read directly from the Smith chart as $z_{IN} = 2 - j1$ or $Z_{IN} = 100 - j50$ Ω.

The magnitude and phase of $\Gamma_0$ are read as indicated in Fig. 2.2.6b. Observe the linear scale for the magnitude of the reflection coefficient. The distance from the origin to $z_L$ can be measured with a ruler or compass and superimposed on the linear scale. The reading of $\Gamma_0$ can be done quite accurately, namely $\Gamma_0 = 0.447 \underline{/63.4°}$.

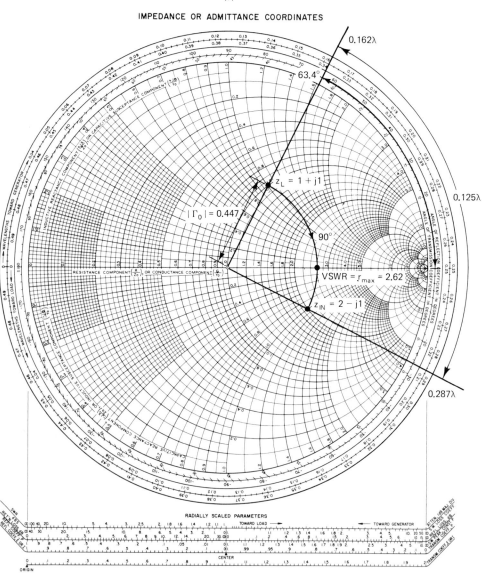

**Figure 2.2.6** Typical transmission-line calculation using the Smith chart.

Sec. 2.3    The Normalized Impedance and Admittance Smith Chart    51

Finally, the VSWR can be calculated from (1.3.11) or the distance from the origin to $z_L$ can be measured and superimposed on the VSWR scale. The value obtained is 2.62. It can also be shown that the value of the maximum resistance in the line is numerically equal to the VSWR. This value is indicated in Fig. 2.2.6b as VSWR = $r_{max} = 2.62$.

## 2.3 THE NORMALIZED IMPEDANCE AND ADMITTANCE SMITH CHART

The conversion of a normalized impedance to a normalized admittance can be done easily in the Smith chart. Since

$$z = \frac{1 + \Gamma}{1 - \Gamma}$$

and

$$y = \frac{1}{z} = \frac{1 - \Gamma}{1 + \Gamma}$$

we observe that rotating $\Gamma$ by $e^{j\pi}$, namely

$$z = \frac{1 + \Gamma e^{j\pi}}{1 - \Gamma e^{j\pi}} = \frac{1 - \Gamma}{1 + \Gamma}$$

results in the value $(1 - \Gamma)/(1 + \Gamma)$, which is identical to the value of the admittance $y$.

**Example 2.3.1**

Find $y$ for $z = 1 + j1$ using the Smith chart.

**Solution.** The graphical solution is illustrated in Fig. 2.3.1. The value of $y$ is read as $0.5 - j0.5$, which of course agrees with

$$y = \frac{1}{z} = \frac{1}{1 + j1} = 0.5 - j0.5$$

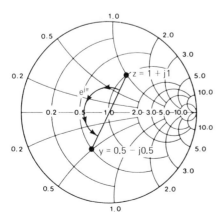

**Figure 2.3.1**   Conversion of $z$ to $y$ in the Smith chart.

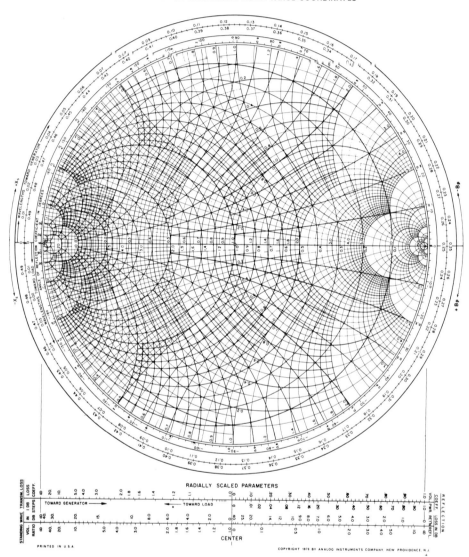

**Figure 2.3.2** The normalized impedance and admittance coordinates Smith chart. (Reproduced with permission of Analog Instruments Co., New Providence, N.J.)

The impedance-to-admittance conversion can also be obtained by rotating the Smith chart by 180° and calling the rotated chart an *admittance chart*. The superposition of the original and the rotated chart is known as the *normal-*

## Sec. 2.3 The Normalized Impedance and Admittance Smith Chart

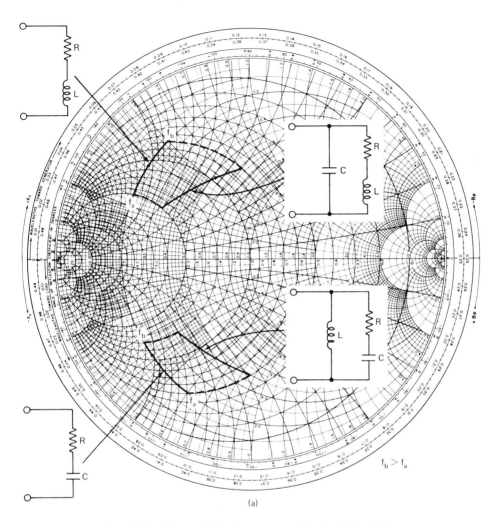

**Figure 2.3.3** Characteristics of some networks in the $ZY$ Smith chart.

*ized impedance and admittance coordinates Smith chart.* We will refer to this Smith chart as the "$ZY$ Smith chart." The $ZY$ Smith chart is shown in Fig. 2.3.2, where the impedance values are shown in red color and the admittance values in green. (See Figure 2.3.2 on inside cover.)

Observe that the upper half of the chart for the admittance coordinates (i.e., green curves) represents normalized admittances having negative susceptances (i.e., $-jb$) and the lower half represents positive susceptances (i.e., $+jb$). The impedance coordinates (i.e., red curves) are the same as in the $Z$ Smith chart.

### Example 2.3.2
Find $y$ for $z = 1 + j1$ using the $ZY$ Smith chart.

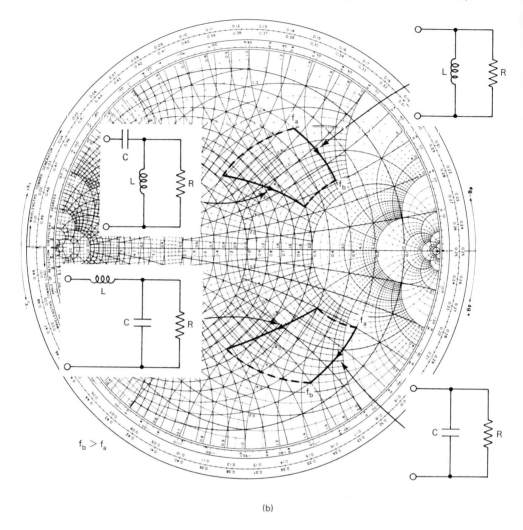

(b)

**Figure 2.3.3** *(continued)*

**Solution.** We can locate in the $ZY$ Smith chart in Fig. 2.3.2 the point $z = 1 + j1$ (red curves), and read directly from the green curves the value $y = 0.5 - j0.5$.

In Section 1.9 some equivalent circuits were obtained from the Smith chart impedance plot of the $S$ parameters. Some practical equivalent circuits are shown in Fig. 2.3.3. Some of these circuits can be used to represent equivalent input and output models of transistors.

## 2.4 IMPEDANCE MATCHING NETWORKS

The need for matching networks arises because amplifiers, in order to deliver maximum power to a load, or to perform in a certain desired way, must be properly terminated at both the input and the output ports. Figure 2.4.1 illustrates a typical situation in which a transistor, in order to deliver maximum power to the 50-Ω load, must have the terminations $Z_s$ and $Z_L$. Although many different types of matching networks can be designed, the eight *ell* sections shown in Fig. 2.4.2 are not only simple to design, but quite practical.

The $ZY$ Smith chart can be used conveniently in the design of matching networks. The effect of adding a series reactance element to an impedance or a parallel susceptance element to an admittance, in the $ZY$ Smith chart, is illustrated in the following example.

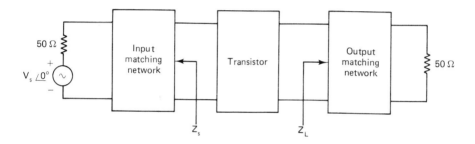

**Figure 2.4.1** Block diagram of a microwave amplifier.

### Example 2.4.1

(a) Illustrate the effect of adding a series inductor $L$ ($z_L = j0.8$) to an impedance $z$ ($z = 0.3 - j0.3$) in the $ZY$ Smith chart.

**Solution.** Figure 2.4.3 shows that the effect of adding a series inductance with $z_L = j0.8$ is to move along a constant-resistance circle from a reactance value of $-0.3$ to a reactance of $0.5$. In other words, the motion is in a clockwise direction along a constant-resistance circle.

(b) Illustrate the effect of adding a series capacitor $C$ ($z_C = -j0.8$) to an impedance $z$ ($z = 0.3 - j0.3$) in the $ZY$ Smith chart.

**Solution.** Figure 2.4.4 shows that the effect of adding a series capacitor with $z_C = -j0.8$ is to move along a constant-resistance circle from a reactance value of $-0.3$ to a reactance of $-1.1$. In other words, the motion is in a counterclockwise direction along a constant-resistance circle.

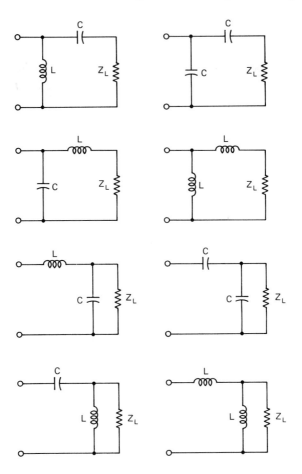

**Figure 2.4.2** Matching networks.

(c) Illustrate the effect of adding a shunt inductor $L$ ($y_L = -j2.4$) to an admittance $y$ ($y = 1.6 + j1.6$) in the $ZY$ Smith chart.

**Solution.** Figure 2.4.5 shows that the effect of adding a shunt inductor with $y_L = -j2.4$ is to move along a constant-conductance circle from a susceptance of 1.6 to a susceptance of $-0.8$. In other words, the motion is in a counterclockwise direction along a constant-conductance circle.

(d) Illustrate the effect of adding a shunt capacitor $C$ ($y_C = j3.4$) to an admittance $y$ ($y = 1.6 + j1.6$) in the $ZY$ Smith chart.

## Sec. 2.4  Impedance Matching Networks

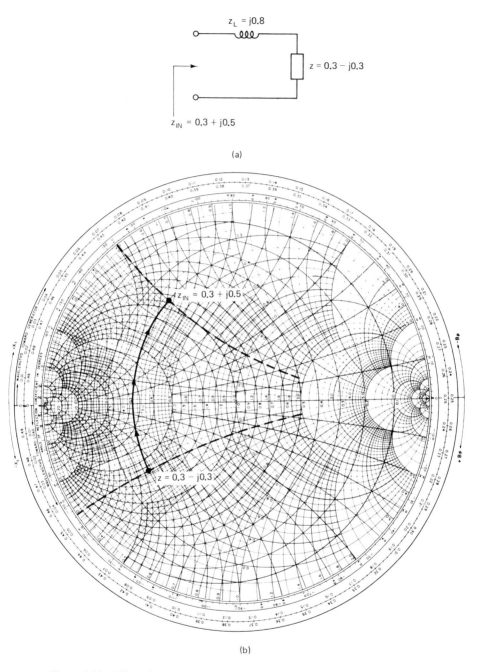

**Figure 2.4.3** Effect of adding a series inductor to an impedance in the $ZY$ Smith chart.

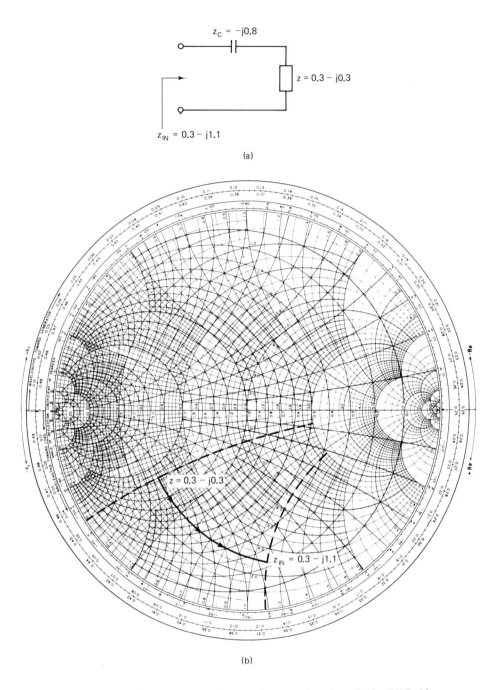

**Figure 2.4.4** Effect of adding a series capacitor to an impedance in the $ZY$ Smith chart.

Sec. 2.4　　Impedance Matching Networks

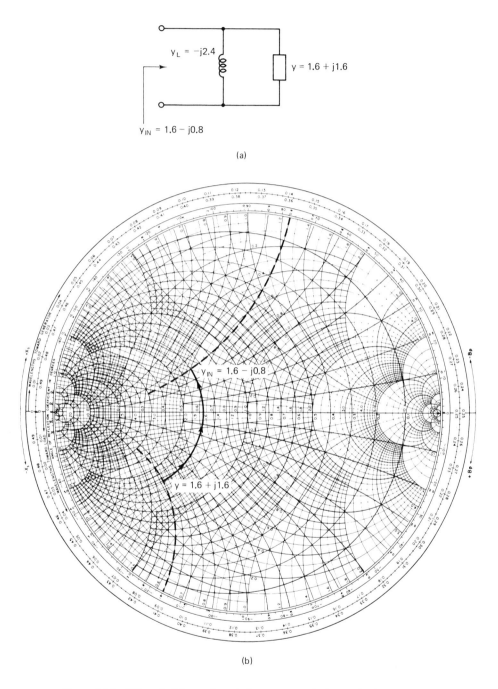

**Figure 2.4.5** Effect of adding a shunt inductor to an admittance in the $ZY$ Smith chart.

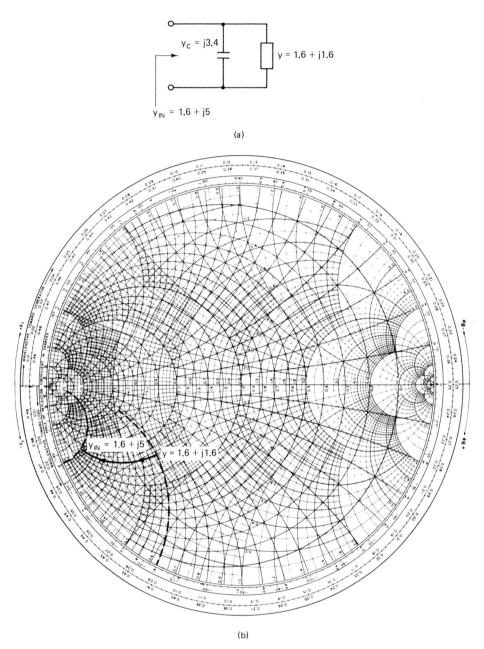

**Figure 2.4.6** Effect of adding a shunt capacitor to an admittance in the $ZY$ Smith chart.

### Sec. 2.4  Impedance Matching Networks

**Solution.**  Figure 2.4.6 shows that the effect of adding a shunt capacitor with $y_C = j3.4$ is to move along a constant-conductance circle from a susceptance of 1.6 to a susceptance of 5. In other words, the motion is in a clockwise direction along a constant-conductance circle.

In conclusion, adding a series reactance produces a motion along a constant-resistance circle in the $ZY$ Smith chart, and adding shunt susceptance produces a motion along a constant-conductance circle in the $ZY$ Smith chart. The four types of motions are illustrated in Fig. 2.4.7.

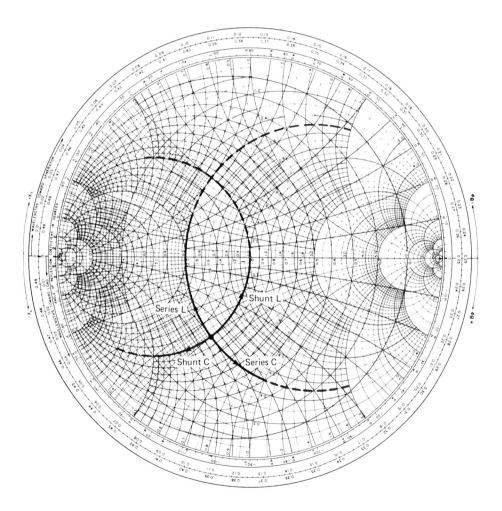

**Figure 2.4.7**  Effect of adding series and shunt elements in the $ZY$ Smith chart.

Designing a matching network in the $ZY$ Smith chart consists of moving along a constant-resistance or constant-conductance circle from one value of impedance or admittance to another. Each motion along a constant-resistance or constant-conductance circle gives the value of an appropriate element. The following examples illustrate the use of the $ZY$ Smith chart in the design of matching networks.

### Example 2.4.2

A load $Z_{LOAD} = 10 + j10$ Ω is to be matched to a 50-Ω line. Design two matching networks and specify the values of $L$ and $C$ at a frequency of 500 MHz.

**Solution.** Selecting the series $L$–shunt $C$ network shown in Fig. 2.4.8a, the matching network is designed as shown in Fig. 2.4.8b. (See Figure 2.4.8b on inside cover.) The motion from point $A$ [i.e., $z_{LOAD} = (10 + j10)/50 = 0.2 + j0.2$] to point $B$ is along a constant-resistance circle, and we obtain for the inductor impedance $z_L = j0.4 - j0.2 = j0.2$. Observe that point $B$ is along the unit constant-conductance circle. The admittance at point $B$ is $y_B = 1 - j2$. The motion from point $B$ to point $C$ (i.e., the origin) is along a constant-conductance circle, and we obtain the capacitor admittance $y_C = 0 - (-j2) = j2$ (or $z_C = 1/j2 = -j0.5$). Therefore, at point $C$, $y_{IN} = z_{IN} = 1$ (or $Z_{IN} = 50$ Ω) and the network is matched to a 50-Ω line. At 500 MHz, the value of $L$ is

$$L = \frac{10}{2\pi(500 \times 10^6)} = 3.18 \text{ nH}$$

and the value of $C$ is

$$C = \frac{1}{25(2\pi)500 \times 10^6} = 12.74 \text{ pF}$$

The matching network at 500 MHz is shown in Fig. 2.4.8c.

The second matching network is shown in Fig. 2.4.9a and the $ZY$ Smith chart design in Fig. 2.4.9b. (See Figure 2.4.9b on inside cover.) The motion from $A$ to $B$ in Fig. 2.4.9b is along a constant-resistance circle; therefore, the impedance of the series capacitor is $z_C = -j0.4 - j0.2 = -j0.6$. The motion from $B$ to $C$ is along a constant-conductance circle; therefore, the admittance of the shunt inductor is $y_L = 0 - j2 = -j2$ (or $z_L = 1/-j2 = j0.5$). The design at 500 MHz is shown in Fig. 2.4.9c.

### Example 2.4.3

Design a matching network to transform a 50-Ω load to the admittance $Y_{OUT} = (8 - j12) \times 10^{-3}$ S.

**Solution.** Figure 2.4.10a illustrates a motion in the $ZY$ Smith chart from the origin (i.e., $z_{LOAD} = 50/50 = 1$) to $y_{OUT} = 50(8 - j12) \times 10^{-3} = 0.4 - j0.6$. (See Figure 2.4.10a on inside cover.) The motion from $A$ to $B$ produces a series capacitor having an impedance of $z_C = -j1.21$. The motion from $B$ to $C$ produces a shunt inductor having an admittance of $y_L = -j0.6 - j0.49 = -j1.09$ (or $z_L = 1/-j1.09 = j0.917$). The matching network is shown in Fig. 2.4.10b.

Sometimes a specific matching network cannot be used to accomplish a given match. For example, any load impedance falling in the marked region in

## Sec. 2.4 Impedance Matching Networks

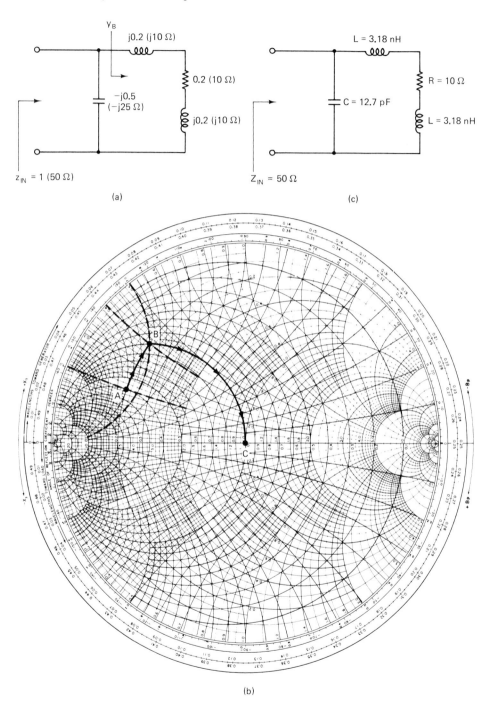

**Figure 2.4.8** Design of a series $L$–shunt $C$ matching network.

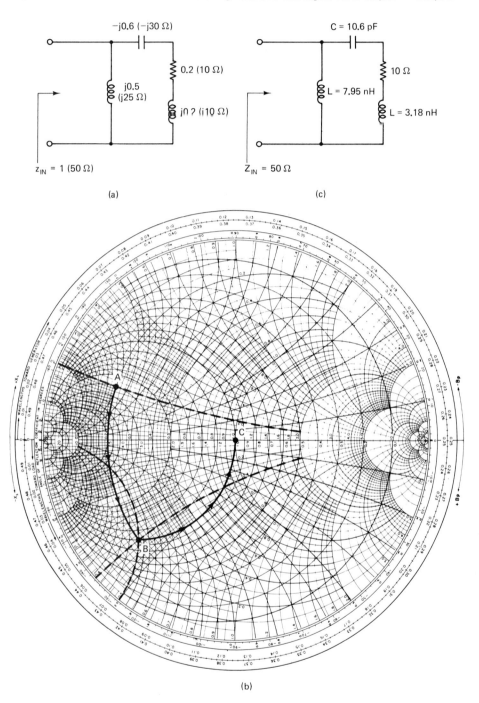

**Figure 2.4.9** Design of a series $C$–shunt $L$ matching network.

Sec. 2.4  Impedance Matching Networks

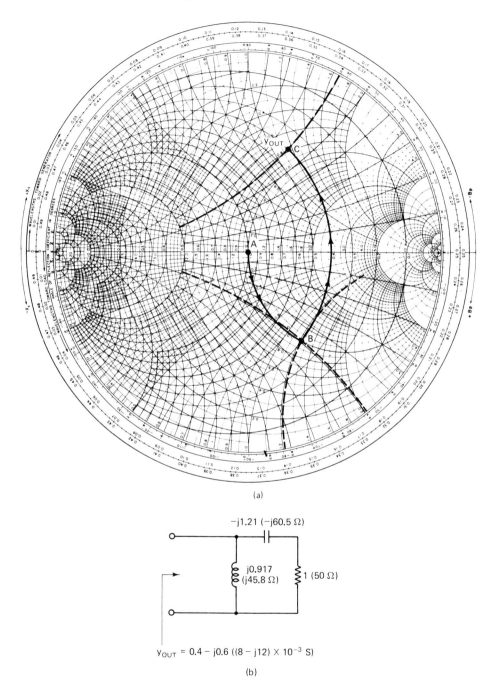

**Figure 2.4.10** Matching a 50-Ω load to a given $y_{OUT}$ using a series $C$–shunt $L$ matching network.

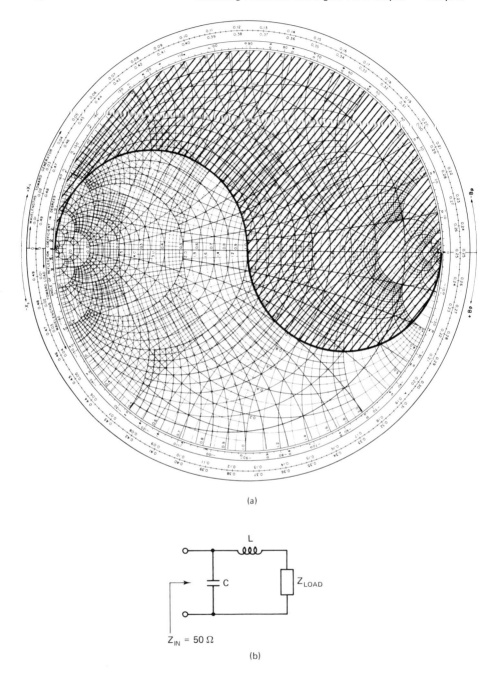

**Figure 2.4.11** Forbidden region in the $ZY$ Smith chart to match a given $Z_{LOAD}$ to 50 Ω using a series $L$–shunt $C$ matching network.

Sec. 2.5    Microstrip Matching Networks    67

Fig. 2.4.11a cannot be matched to 50 Ω with the network in Fig. 2.4.11b because adding a series $L$ produces motion, in a clockwise direction, away from any constant-conductance circle that passes through the origin.

## 2.5 MICROSTRIP MATCHING NETWORKS

Microstrip lines are used extensively in building microwave transistor amplifiers because they are easily fabricated using printed-circuit techniques. Network interconnections and the placement of lumped and transistor devices are easily made on its metal surface. The superior performance characteristics of the microstrip line makes it one of the most important medium of transmission in microwave transistor amplifiers and in microwave integrated-circuit technology.

A microstrip line is by definition a transmission line consisting of a strip conductor and a ground plane separated by a dielectric medium. Figure 2.5.1 illustrates the microstrip geometry. The dielectric material serves as a substrate and it is sandwiched between the strip conductor and the ground plane. Some typical dielectric substrates are: Duroid, a trademark of Rogers Corporation (Chandler, Arizona), ($\varepsilon = 2.56\varepsilon_0$), quartz ($\varepsilon = 3.78\varepsilon_0$), alumina ($\varepsilon = 9.7\varepsilon_0$), and silicon ($\varepsilon = 11.7\varepsilon_0$). The relative dielectric constant of the substrate, $\varepsilon_r$, follows from $\varepsilon = \varepsilon_r \varepsilon_0$, where $\varepsilon_0 = 8.854 \times 10^{-12}$ F/m.

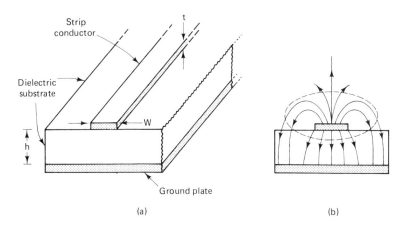

**Figure 2.5.1** Microstrip geometry and field configuration. In (b), the solid lines represent electric field lines and the dashed line represents a magnetic field line.

The electromagnetic field lines in the microstrip are not contained entirely in the substrate. Therefore, the propagating mode in the microstrip is not a pure transverse electromagnetic mode (TEM mode) but a quasi-TEM. Assuming a quasi-TEM mode of propagation in the microstrip line, the phase velocity is given by

$$v_p = \frac{c}{\sqrt{\varepsilon_{ff}}} \qquad (2.5.1)$$

where $c$ is the speed of light (i.e., $3 \times 10^8$ m/s) and $\varepsilon_{ff}$ is the effective relative dielectric constant of the dielectric substrate. The effective relative dielectric constant of the microstrip is related to the relative dielectric constant of the dielectric substrate, and also takes into account the effect of the external electromagnetic fields.

The characteristic impedance of the microstrip line is given by

$$Z_o = \frac{1}{v_p C} \qquad (2.5.2)$$

where $C$ is the capacitance per unit length of the microstrip. The wavelength in the microstrip line is given by

$$\lambda = \frac{v_p}{f} = \frac{c}{f\sqrt{\varepsilon_{ff}}} = \frac{\lambda_0}{\sqrt{\varepsilon_{ff}}} \qquad (2.5.3)$$

where $\lambda_0$ is the free-space wavelength.

As seen from (2.5.1), (2.5.2), and (2.5.3), the evaluation of $v_p$, $Z_o$, and $\lambda$ in a microstrip line requires the evaluation of $\varepsilon_{ff}$ and $C$. There are different methods for determining $\varepsilon_{ff}$ and $C$ and, of course, closed-form expressions are of great importance in microstrip-line design. The evaluation of $\varepsilon_{ff}$ and $C$ based on a quasi-TEM mode is accurate for design purposes at lower microwave frequencies. However, at higher microwave frequencies the longitudinal components of the electromagnetic fields are significant and the quasi-TEM assumption is no longer valid.

A useful set of relations for the characteristic impedance, assuming zero or negligible thickness of the strip conductor (i.e., $t/h < 0.005$), are [2.1]:

*For $W/h \leq 1$:*

$$Z_o = \frac{60}{\sqrt{\varepsilon_{ff}}} \ln\left(8\frac{h}{W} + 0.25\frac{W}{h}\right) \qquad (2.5.4)$$

where

$$\varepsilon_{ff} = \frac{\varepsilon_r + 1}{2} + \frac{\varepsilon_r - 1}{2}\left[\left(1 + 12\frac{h}{W}\right)^{-1/2} + 0.04\left(1 - \frac{W}{h}\right)^2\right] \qquad (2.5.5)$$

*For $W/h \geq 1$:*

$$Z_o = \frac{120\pi/\sqrt{\varepsilon_{ff}}}{W/h + 1.393 + 0.667 \ln(W/h + 1.444)} \qquad (2.5.6)$$

## Sec. 2.5  Microstrip Matching Networks

where

$$\varepsilon_{ff} = \frac{\varepsilon_r + 1}{2} + \frac{\varepsilon_r - 1}{2}\left(1 + 12\frac{h}{W}\right)^{-1/2} \quad (2.5.7)$$

Plots of the characteristic impedance, as well as the normalized wavelength, as a function of $W/h$ are shown in Figs. 2.5.2 and 2.5.3.

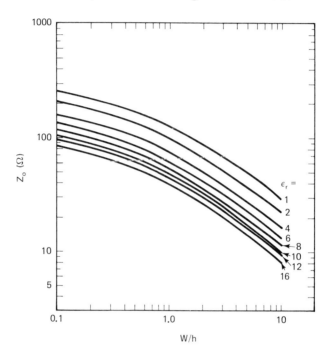

**Figure 2.5.2** Characteristic impedance of the microstrip line versus $W/h$. (From H. Sobol [2.2]; copyright 1971, IEEE; reproduced with permission of IEEE.)

Based on the results in (2.5.3), (2.5.5), and (2.5.7) and/or in experimental data, the wavelength in the microstrip line, assuming zero or negligible thickness (i.e., $t/h \leq 0.005$) for the strip conductor, is given by the relations [2.3]:

*For $W/h \geq 0.6$:*

$$\lambda = \frac{\lambda_0}{\sqrt{\varepsilon_r}}\left[\frac{\varepsilon_r}{1 + 0.63(\varepsilon_r - 1)(W/h)^{0.1255}}\right]^{1/2} \quad (2.5.8)$$

*For $W/h < 0.6$:*

$$\lambda = \frac{\lambda_0}{\sqrt{\varepsilon_r}}\left[\frac{\varepsilon_r}{1 + 0.6(\varepsilon_r - 1)(W/h)^{0.0297}}\right]^{1/2} \quad (2.5.9)$$

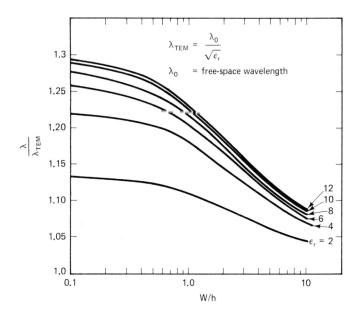

**Figure 2.5.3** Normalized wavelength of the microstrip line versus $W/h$. (From H. Sobol [2.2]; copyright 1971, IEEE; reproduced with permission of IEEE.)

For design purposes a set of equations relating $Z_o$ and $\varepsilon_r$ to the ratio $W/h$ of the microstrip line is desirable. Assuming zero or negligible thickness of the strip conductor (i.e., $t/h \leq 0.005$), the expressions are [2.2]:

*For $W/h \leq 2$:*

$$\frac{W}{h} = \frac{8e^A}{e^{2A} - 2} \qquad (2.5.10)$$

*For $W/h \geq 2$:*

$$\frac{W}{h} = \frac{2}{\pi}\left\{B - 1 - \ln(2B - 1) + \frac{\varepsilon_r - 1}{2\varepsilon_r}\left[\ln(B - 1) + 0.39 - \frac{0.61}{\varepsilon_r}\right]\right\}$$

$$(2.5.11)$$

where

$$A = \frac{Z_o}{60}\sqrt{\frac{\varepsilon_r + 1}{2}} + \frac{\varepsilon_r - 1}{\varepsilon_r + 1}\left(0.23 + \frac{0.11}{\varepsilon_r}\right)$$

and

$$B = \frac{377\pi}{2Z_o\sqrt{\varepsilon_r}}$$

## Sec. 2.5 Microstrip Matching Networks

The zero or negligible thickness formulas given in (2.5.4) to (2.5.11) can be modified to include the thickness of the strip conductor. The first-order effect of a strip conductor of finite thickness $t$ is to increase the capacitance. Therefore, an approximate correction is made by replacing the strip width $W$ by the effective width $W_{\text{eff}}$. The following relation for $W_{\text{eff}}/h$ are useful when $t < h$ and $t < W/2$:

For $W/h \geq 1/2\pi$:

$$\frac{W_{\text{eff}}}{h} = \frac{W}{h} + \frac{t}{\pi h}\left(1 + \ln\frac{2h}{t}\right)$$

For $W/h \leq 1/2\pi$:

$$\frac{W_{\text{eff}}}{h} = \frac{W}{h} + \frac{t}{\pi h}\left(1 + \ln\frac{4\pi W}{t}\right)$$

The restrictions $t < h$ and $t < W/2$ are usually satisfied since for dielectric substrates a typical thickness is $t = 0.002$ in.

The formulas presented thus far are valid at frequencies where the quasi-TEM assumption can be made. When the quasi-TEM assumption is not valid, $\varepsilon_{ff}$ and $Z_o$ are functions of frequency and, therefore, the microstrip line becomes dispersive. The phase velocity of the microstrip line decreases with increasing frequency. Therefore, $\varepsilon_{ff}(f)$ increases with frequency. Also, the characteristic impedance of the microstrip line increases with frequency, and it follows that the effective width $W_{\text{eff}}(f)$ decreases.

The frequency below which dispersion may be neglected is given by

$$f_o(\text{GHz}) = 0.3\sqrt{\frac{Z_o}{h\sqrt{\varepsilon_r - 1}}}$$

where $h$ must be expressed in centimeters.

An analytical expression that shows the effect of dispersion in $\varepsilon_{ff}(f)$ is [2.1]

$$\varepsilon_{ff}(f) = \varepsilon_r - \frac{\varepsilon_r - \varepsilon_{ff}}{1 + G(f/f_p)^2} \quad (f \text{ in GHz})$$

where

$$f_p = \frac{Z_o}{8\pi h} \quad (h \text{ in cm})$$

and

$$G = 0.6 + 0.009 Z_o$$

Observe that when $f_p \gg f$, then $\varepsilon_{ff}(f) \simeq \varepsilon_{ff}$. In other words, high-impedance lines on thin substrates are less dispersive.

The expression for the dispersion in $Z_o$ is [2.1]

$$Z_o(f) = \frac{377h}{W_{eff}(f)\sqrt{\varepsilon_{ff}}}$$

where

$$W_{eff}(f) = W + \frac{W_{eff}(0) - W}{1 + (f/f_p)^2}$$

and

$$W_{eff}(0) = \frac{377h}{Z_o(0)\sqrt{\varepsilon_{ff}(0)}}$$

Another characteristic of the microstrip line is its attenuation. The attenuation constant is a function of the microstrip geometry, the electrical properties of the dielectric substrate and the conductors, and the frequency.

There are two types of losses in a microstrip line: a dielectric substrate loss and the ohmic skin loss in the conductors. The losses can be expressed as a loss per unit length along the microstrip line in terms of the attenuation factor $\alpha$. Since the power carried by a wave traveling in the positive direction in a quasi-TEM mode is given by

$$P^+(z) = \frac{1}{2}(V^+ e^{-\alpha z} I^+ e^{-\alpha z}) = \frac{1}{2}\frac{|V^+|^2}{Z_o} e^{-2\alpha z}$$

$$= P_0 e^{-2\alpha z} \qquad (2.5.12)$$

where $P_0 = |V^+|^2/2Z_o$ is the power at $z = 0$. Then, from (2.5.12) we can write

$$\alpha = \frac{-dP(z)/dz}{2P(z)} = \alpha_d + \alpha_c$$

where $\alpha_d$ is the dielectric loss factor and $\alpha_c$ the conduction loss factor.

A useful set of expressions for calculating $\alpha_d$ is [2.1]:

*For a dielectric with low losses:*

$$\alpha_d = 27.3 \frac{\varepsilon_r}{\sqrt{\varepsilon_{ff}}} \frac{\varepsilon_{ff} - 1}{\varepsilon_r - 1} \frac{\tan \delta}{\lambda_0} \quad \frac{dB}{cm} \qquad (2.5.13)$$

where the loss tangent $\delta$ is given by

$$\tan \delta = \frac{\sigma}{\omega \varepsilon}$$

*For a dielectric with $\sigma \neq 0$:*

$$\alpha_d = 4.34 \frac{\varepsilon_{ff} - 1}{\sqrt{\varepsilon_{ff}}(\varepsilon_r - 1)} \left(\frac{\mu_0}{\varepsilon_0}\right)^{1/2} \sigma \quad \frac{dB}{cm} \qquad (2.5.14)$$

Sec. 2.5   Microstrip Matching Networks

In (2.5.13) and (2.5.14), $\sigma$ is the conductivity of the dielectric and $\mu_0 = 4\pi \times 10^{-7}$ H/m.

A set of expressions for calculating $\alpha_c$ is [2.1]:

For $W/h \to \infty$:

$$\alpha_c = \frac{8.68}{Z_o W} R_s$$

where

$$R_s = \sqrt{\frac{\pi f \mu_0}{\sigma}}$$

For $W/h \le 1/2\pi$:

$$\alpha_c = \frac{8.68 R_s P}{2\pi Z_o h}\left[1 + \frac{h}{W_{\text{eff}}} + \frac{h}{\pi W_{\text{eff}}}\left(\ln\frac{4\pi W}{t} + \frac{t}{W}\right)\right]$$

For $\frac{1}{2}\pi < W/h \le 2$:

$$\alpha_c = \frac{8.68 R_s}{2\pi Z_o h} PQ$$

For $W/h \ge 2$:

$$\alpha_c = \frac{8.68 R_s Q}{Z_o h}\left\{\frac{W_{\text{eff}}}{h} + \frac{2}{\pi}\ln\left[2\pi e\left(\frac{W_{\text{eff}}}{2h} + 0.94\right)\right]\right\}^{-2}\left[\frac{W_{\text{eff}}}{h} + \frac{W_{\text{eff}}/\pi h}{(W_{\text{eff}}/2h) + 0.94}\right]$$

where

$$P = 1 - \left(\frac{W_{\text{eff}}}{4h}\right)^2$$

and

$$Q = 1 + \frac{h}{W_{\text{eff}}} + \frac{h}{\pi W_{\text{eff}}}\left(\ln\frac{2h}{t} - \frac{t}{h}\right)$$

In dielectric substrates, the dielectric losses are normally smaller than conductor losses. However, dielectric losses in silicon substrates can be of the same order or larger than conductor losses.

The quality factor $Q$ of a microstrip line is calculated from

$$Q = \frac{\beta}{2\alpha}$$

where

$$\beta = \frac{2\pi}{\lambda}$$

and $\alpha$ is the total loss. Therefore,

$$Q = \frac{\pi}{\lambda \alpha}$$

or in decibels we can write

$$Q = \frac{8.686\pi}{\lambda \alpha} \quad \text{dB}$$

$$= \frac{27.3}{\alpha} \quad \frac{\text{dB}}{\lambda}$$

where we used the fact that 1 dB = 8.686 nepers.

A microstrip line also has radiation losses. The effect of radiation losses can be accounted for in terms of the radiation quality factor $Q_r$ given by [2.1]

$$Q_r = \frac{Z_o}{480\pi(h/\lambda_0)F}$$

where

$$F = \frac{\varepsilon_{ff}(f) + 1}{\varepsilon_{ff}(f)} - \frac{(\varepsilon_{ff}(f) - 1)^2}{2[\varepsilon_{ff}(f)]^{2/3}} \ln \frac{\sqrt{\varepsilon_{ff}(f)} + 1}{\sqrt{\varepsilon_{ff}(f)} - 1}$$

is known as the *radiation factor*.

The total $Q$, called $Q_T$, of a microstrip resonator can be expressed as

$$\frac{1}{Q_T} = \frac{1}{Q_c} + \frac{1}{Q_d} + \frac{1}{Q_r}$$

where $Q_d$ and $Q_c$ are the quality factors of the dielectric (i.e., $Q_d = \pi/\lambda\alpha_d$) and conductor (i.e., $Q_c = \pi/\lambda\alpha_c$), respectively.

The impedance transforming properties of transmission lines can be used in the design of matching networks. A microstrip line can be used as a series transmission line, as an open-circuited stub or as a short-circuited stub. In fact, a series microstrip line together with a short- or open-circuited shunt stub can transform a 50-$\Omega$ resistor into any value of impedance. Also, a quarter-wave microstrip line can be used to change a 50-$\Omega$ resistor to any value of resistance. This line, together with a short- or open-circuited shunt stub, can be used to transform 50 $\Omega$ to any value of impedance. The following example illustrates the use of microstrip lines in matching networks.

**Example 2.5.1**

Design two microstrip matching networks for the amplifier shown in Fig. 2.5.4, whose reflection coefficients for a good match, in a 50-$\Omega$ system, are $\Gamma_s = 0.614 \underline{|160°}$ and $\Gamma_L = 0.682 \underline{|97°}$.

**Solution.** *Design 1:* The amplifier block diagram is shown in Fig. 2.5.4. The normalized impedances and admittances associated with $\Gamma_s$ and $\Gamma_L$ can be read, to reasonable

## Sec. 2.5  Microstrip Matching Networks

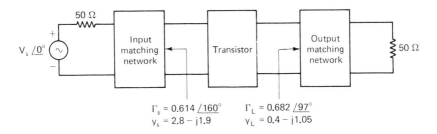

**Figure 2.5.4**  Amplifier block diagram.

accuracy, from the $ZY$ Smith chart, namely

$$y_s = \frac{1}{z_s} = \frac{1}{0.245 + j0.165} = 2.8 - j1.9$$

and

$$y_L = \frac{1}{z_L} = \frac{1}{0.325 + j0.83} = 0.4 - j1.05$$

In order to design the input matching network, we locate $y_s$ in the $Y$ Smith chart shown in Fig. 2.5.5a. The shortest length of microstrip line plus stub is obtained by using an open-circuited shunt stub of length $0.159\lambda$ to move from the origin (i.e., 50 Ω) to point $A$ on the Smith chart, and then using a transmission line length of $0.099\lambda$ to move from $A$ to $y_s$.

Next, we locate $y_L$ in Fig. 2.5.5b and follow a similar procedure. In this case, the shortest length of microstrip line plus stub is obtained by using a short-circuited shunt stub of length $0.077\lambda$ to move from the origin to point $B$. Then a series transmission line of length $0.051\lambda$ to move from $B$ to $y_L$.

The complete design showing the transistor, the microstrip matching network, and the dc supply is shown in Fig. 2.5.6. The characteristic impedance of all microstrip lines is 50 Ω.

The capacitors $C_A$ are coupling capacitors. Typical values for the chip capacitors $C_A$ are 200 to 1000 pF, high-$Q$ capacitor. The bypass capacitors $C_B$ (i.e., chip capacitors, 50 to 500 pF) provide the ac short circuits for the $0.077\lambda$ and $\lambda/4$ short-circuited stubs. The $\lambda/4$ short-circuited stub, high-impedance line (denoted by $Z_o \gg$), provides the dc path for the base supply voltage $V_{BB}$. It also presents an open circuit to the ac signal at the base of the transistor. The narrowest practical line (i.e., large $Z_o$) should be used for the $\lambda/4$ short-circuited stub to avoid unwanted ac coupling.

To minimize transition interaction between the shunt stubs and the series transmission lines, the shunt stubs are usually balanced along the series transmission line. A schematic of the amplifier using balanced shunt stubs is shown in Fig. 2.5.7. The schematic also shows that 50-Ω lines were added on both sides of $C_A$ to provide a soldering area.

In Fig. 2.5.7 two parallel shunt stubs must provide the same admittance as the single stub in Fig. 2.5.6. Therefore, the admittance of each side of the balanced stub must be equal to half of the total admittance. For example, each side of the input

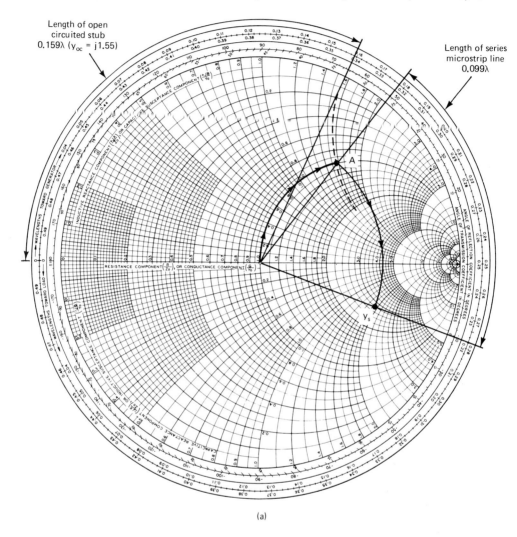

**Figure 2.5.5** (a) Input matching network design; (b) output matching network design.

balanced shunt stubs must have an admittance of $y = j1.55/2 = j0.775$. Using the Smith chart, we obtain that the length of each side must be $0.105\lambda$. Observe that the length of the shunt stubs in Fig. 2.5.7 is not equal to the total length of the balance stubs in Fig. 2.5.6. Of course, a simple check will show that the admittance seen by the series transmission line is the same in both cases.

If we use Duroid ($\varepsilon_r = 2.23$, $h = 0.7874$ mm) to build the amplifier, we find from the results in Section 2.5 that a characteristic impedance of 50 Ω is obtained with $W = 2.42$ mm and $\varepsilon_{ff} = 1.91$. The microstrip wavelength in the 50-Ω Duroid microstrip line is $\lambda = \lambda_0/\sqrt{1.91} = 0.7236\lambda_0$, where $\lambda_0 = 30$ cm at $f = 1$ GHz. For a

## Sec. 2.5  Microstrip Matching Networks

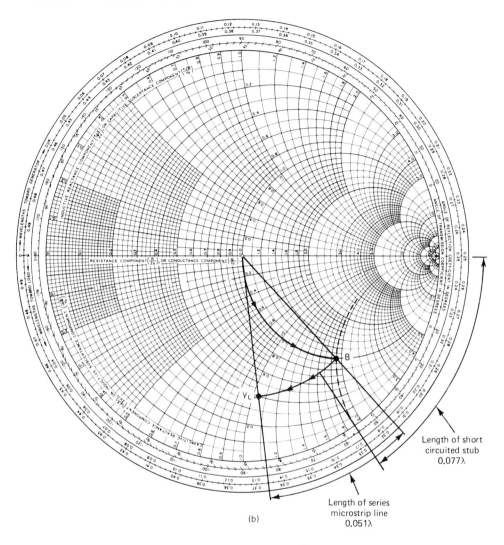

(b)

**Figure 2.5.5** (*continued*)

characteristic impedance of 100 Ω in the $\lambda/4$ line, the width must be $W = 0.7$ mm. The line lengths in Fig. 2.5.7 are

$$0.105\lambda = 2.28 \text{ cm}$$

$$0.099\lambda = 2.15 \text{ cm}$$

$$0.051\lambda = 1.10 \text{ cm}$$

$$0.129\lambda = 2.80 \text{ cm}$$

$$\lambda/4 = 5.43 \text{ cm}$$

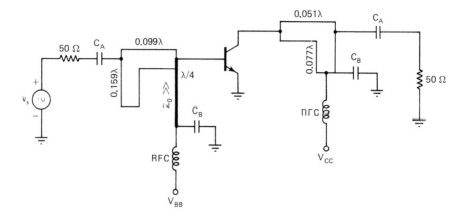

**Figure 2.5.6** Complete amplifier schematic. The characteristic impedance of the microstrip lines is 50 Ω.

*Design 2:* This method uses microstrip lines with different characteristic impedances. The design requires the transformation of 50 Ω to $Y_s = (2.8 - j1.9)/50 = 0.056 - j0.038$ S. A quarter-wave transformer can be used to transform the source impedance of 50 Ω to the resistance $1/0.056 = 17.86$ Ω. The characteristic impedance of the quarter-wave transformer is

$$Z_o = \sqrt{50(17.86)} = 29.9 \text{ Ω}$$

An open-circuited stub provides an admittance of $Y_{oc} = jY_o \tan \beta l$. Therefore, an open-circuited shunt stub of length $3\lambda/8$, or a short-circuited shunt stub of length $\lambda/8$, looks like a shunt inductor having the admittance $-jY_o$. Using an open-circuited shunt stub of length $3\lambda/8$, we find its characteristic impedance to be $Z_o = 1/Y_o = 1/0.038 = 26.32$ Ω.

Similarly, for the output matching network $[Y_L = (0.4 - j1.05)/50 = 0.008 - j0.021$ S] a quarter-wave line of characteristic impedance

$$Z_o = \sqrt{50(125)} = 79.1 \text{ Ω}$$

transforms the 50-Ω load to a resistance of value $1/0.008 = 125$ Ω. An open-circuited shunt stub of length $3\lambda/8$ and characteristic impedance $Z_o = 1/Y_o = 1/0.021 = 47.6$ Ω produces the required admittance of $-j0.021$.

The complete amplifier is shown in Fig. 2.5.8a. Figure 2.5.8b shows the amplifier using balanced shunt stubs of length $3\lambda/8$ to minimize the microstrip transition interaction. Observe that in the balance stubs the lengths were kept at $3\lambda/8$, but the characteristic impedance was doubled so that the parallel combinations produce the required 26.32 Ω and 47.6 Ω.

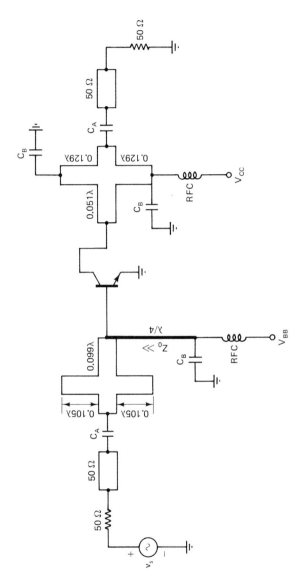

**Figure 2.5.7** Complete amplifier schematic using balanced shunt stubs. The characteristic impedance of the microstrip lines is 50 Ω.

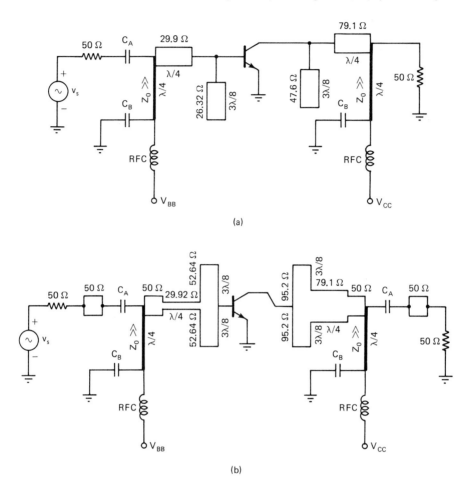

**Figure 2.5.8** Matching network design using microstrip lines with different characteristic impedances.

## 2.6 SIGNAL FLOW GRAPH

A signal flow graph is a convenient technique to represent and analyze the transmission and reflection of waves in a microwave amplifier. Once the signal flow graph is developed, relations between the variables can be obtained using Mason's rule. The flow graph technique permits expressions, such as power gains and voltage gains of complex microwave amplifiers, to be easily derived. Certain rules are followed in constructing a signal flow graph:

1. Each variable is designated as a node.
2. The $S$ parameters and reflection coefficients are represented by branches.

3. Branches enter dependent variable nodes and emanate from independent variable nodes. The independent variable nodes are the incident waves and the reflected waves are dependent variable nodes.
4. A node is equal to the sum of the branches entering it.

The signal flow graph of the $S$ parameters of a two-port network is shown in Fig. 2.6.1. Observe that $b_1$ and $b_2$ are the dependent nodes and $a_1$ and $a_2$ the independent nodes. The complete signal flow graph of the two-port network is shown in Fig. 2.6.2.

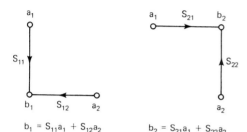

Figure 2.6.1 Signal flow graph for the scattering parameter equations.

The signal flow graph in Fig. 2.6.2 shows the relationship between the traveling waves. The incident wave $a_1$ at port 1 gets partly transmitted (i.e., $S_{21}a_1$) to become part of $b_2$, and partly reflected (i.e., $S_{11}a_1$) to become part of $b_1$. Similarly, the incident wave $a_2$ at port 2 gets partly transmitted (i.e., $S_{12}a_2$) to become part of $b_1$ and partly reflected (i.e., $S_{22}a_2$) to become part of $b_2$.

In order to obtain the signal flow graph of a microwave amplifier we need to obtain the signal flow graph of a signal generator with some internal impedance, and the signal flow graph of a load impedance.

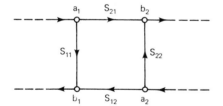

Figure 2.6.2 Signal flow graph of a two-port network.

Figure 2.6.3a shows a voltage-source generator with impedance $Z_s$. At the terminals we can write

$$V_g = E_s + I_g Z_s \tag{2.6.1}$$

Using (1.4.1) and (1.4.2), we can express (2.6.1) in terms of traveling waves, namely

$$V_g^+ + V_g^- = E_s + \left(\frac{V_g^+}{Z_o} - \frac{V_g^-}{Z_o}\right) Z_s$$

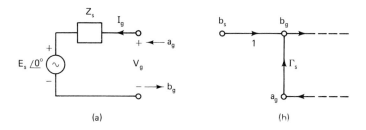

**Figure 2.6.3** Signal flow graph of a voltage-source generator.

Solving for $V_g^-$, we obtain

$$b_g = b_s + \Gamma_s a_g \qquad (2.6.2)$$

where

$$b_g = \frac{V_g^-}{\sqrt{Z_o}}$$

$$a_g = \frac{V_g^+}{\sqrt{Z_o}}$$

$$b_s = \frac{E_s \sqrt{Z_o}}{Z_s + Z_o}$$

and

$$\Gamma_s = \frac{Z_s - Z_o}{Z_s + Z_o}$$

From (2.6.2) the signal flow graph in Fig. 2.6.3b follows.

For the load impedance shown in Fig. 2.6.4a we can write

$$V_L = Z_L I_L$$

In terms of traveling waves, we obtain

$$V_L^+ + V_L^- = Z_L \left( \frac{V_L^+}{Z_o} - \frac{V_L^-}{Z_o} \right)$$

or

$$b_L = \Gamma_L a_L \qquad (2.6.3)$$

where

$$b_L = \frac{V_L^-}{\sqrt{Z_o}}$$

$$a_L = \frac{V_L^+}{\sqrt{Z_o}}$$

Sec. 2.6    Signal Flow Graph

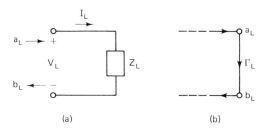

**Figure 2.6.4** Signal flow graph of a load impedance.

and

$$\Gamma_L = \frac{Z_L - Z_o}{Z_L + Z_o}$$

The signal flow graph follows from (2.6.3) and it is shown in Fig. 2.6.4b.

We can now combine the signal flow graph for the two-port network in Fig. 2.6.2 with the signal flow graphs of the signal generator (i.e., Fig. 2.6.3b) and the load (i.e., Fig. 2.6.4b). Observe that the nodes $b_g$, $a_g$, $b_L$, and $a_L$ are identical to $a_1$, $b_1$, $a_2$, and $b_2$, respectively. The resulting signal flow graph of a microwave amplifier is shown in Fig. 2.6.5.

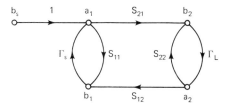

**Figure 2.6.5** Signal flow graph of a microwave amplifier.

To determine the ratio or transfer function $T$ of a dependent to an independent variable, we apply Mason's rule, namely

$$T = \frac{P_1[1 - \sum L(1)^{(1)} + \sum L(2)^{(1)} - \cdots] + P_2[1 - \sum L(1)^{(2)} + \cdots] + \cdots}{1 - \sum L(1) + \sum L(2) - \sum L(3) + \cdots}$$

where the different terms are defined as follows.

The terms $P_1$, $P_2$, and so on, are the different paths connecting the dependent and independent variables whose transfer function $T$ is to be determined. A path is defined as a set of consecutive, codirectional branches along which no node is encountered more than once as we move in the graph from the independent to the dependent node. The value of the path is the product of all branch coefficients along the path. For example, in Fig. 2.6.5 $b_s$ is the only independent variable. To determine the ratio $b_1/b_s$ we identify two paths, $P_1 = S_{11}$ and $P_2 = S_{21} \Gamma_L S_{12}$.

The term $\sum L(1)$ is the sum of all first-order loops. A first-order loop is defined as the product of the branches encountered in a round trip as we move from a node in the direction of the arrows back to that original node. In Fig. 2.6.5, $S_{11}\Gamma_s$, $S_{21}\Gamma_L S_{12}\Gamma_s$, and $S_{22}\Gamma_L$ are first-order loops.

The term $\sum L(2)$ is the sum of all second-order loops. A second-order loop is defined as the product of any two nontouching first-order loops. In Fig. 2.6.5, $S_{11}\Gamma_s$ and $S_{22}\Gamma_L$ do not touch; therefore, the product $S_{11}\Gamma_s S_{22}\Gamma_L$ is a second-order loop.

The term $\sum L(3)$ is the sum of all third-order loops. A third-order loop is defined as the product of three nontouching first-order loops. In Fig. 2.6.5 there are no third-order loops. Of course, the terms $\sum L(4)$, $\sum L(5)$, and so on, represent fourth-, fifth-, and higher-order loops.

The terms $\sum L(1)^{(P)}$ is the sum of all first-order loops that do not touch the path $P$ between the independent and dependent variables. In Fig. 2.6.5, for the path $P_1 = S_{11}$ we find that $\sum L(1)^{(1)} = \Gamma_L S_{22}$, and for the path $P_2 = S_{21}\Gamma_L S_{12}$ we find that $\sum L(1)^{(2)} = 0$.

The term $\sum L(2)^{(P)}$ is the sum of all second-order loops that do not touch the path $P$ between the independent and dependent variables. In Fig. 2.6.5, we find that $\sum L(2)^{(P)} = 0$. Of course, $\sum L(3)^{(P)}$, $\sum L(4)^{(P)}$, and so on, represent higher-order loops that do not touch the path $P$.

For the transfer function $b_1/b_s$ in Fig. 2.6.5 we have found that $P_1 = S_{11}$, $P_2 = S_{21}\Gamma_L S_{12}$, $\sum L(1) = S_{11}\Gamma_s + S_{22}\Gamma_L + S_{21}\Gamma_L S_{12}\Gamma_s$, $\sum L(2) = S_{11}\Gamma_s S_{22}\Gamma_L$, and $\sum L(1)^{(1)} = \Gamma_L S_{22}$. Therefore, using Mason's rule, we obtain

$$\frac{b_1}{b_s} = \frac{S_{11}(1 - \Gamma_L S_{22}) + S_{21}\Gamma_L S_{12}}{1 - (S_{11}\Gamma_s + S_{22}\Gamma_L + S_{21}\Gamma_L S_{12}\Gamma_s) + S_{11}\Gamma_s S_{22}\Gamma_L}$$

## 2.7 APPLICATIONS OF SIGNAL FLOW GRAPHS

The first application of signal flow graph analysis is in the calculation of the input reflection coefficient, called $\Gamma_{IN}$, when a load is connected to the output of a two-port network. The signal flow graph is shown in Fig. 2.7.1.

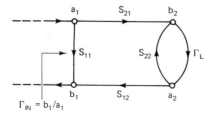

**Figure 2.7.1** Signal flow graph for the input reflection coefficient $\Gamma_{IN}$.

## Sec. 2.7 Applications of Signal Flow Graphs

The input reflection coefficient $\Gamma_{IN}$ is defined as

$$\Gamma_{IN} = \frac{b_1}{a_1}$$

Observing that $P_1 = S_{11}$, $P_2 = S_{21}\Gamma_L S_{12}$, $\sum L(1) = S_{22}\Gamma_L$, and $\sum L(1)^{(1)} = S_{22}\Gamma_L$, we can use Mason's rule to obtain

$$\Gamma_{IN} = \frac{S_{11}(1 - S_{22}\Gamma_L) + S_{21}\Gamma_L S_{12}}{1 - S_{22}\Gamma_L}$$

$$= S_{11} + \frac{S_{12}S_{21}\Gamma_L}{1 - S_{22}\Gamma_L} \qquad (2.7.1)$$

If $\Gamma_L = 0$, it follows from (2.7.1) that $\Gamma_{IN} = S_{11}$. Also, when there is no transmission from the output to the input (i.e., when $S_{12} = 0$), it follows that $\Gamma_{IN} = S_{11}$. When $S_{12} = 0$, we call the device represented by the two-port a unilateral device.

Similarly, we can calculate the output reflection coefficient $\Gamma_{OUT} = b_2/a_2$ with $b_s = 0$ from the signal flow graph shown in Fig. 2.7.2. The expression for $\Gamma_{OUT}$ is

$$\Gamma_{OUT} = S_{22} + \frac{S_{12}S_{21}\Gamma_s}{1 - S_{11}\Gamma_s} \qquad (2.7.2)$$

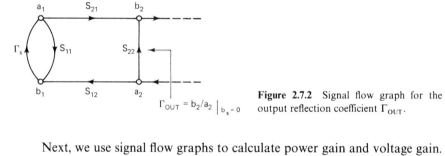

Figure 2.7.2 Signal flow graph for the output reflection coefficient $\Gamma_{OUT}$.

Next, we use signal flow graphs to calculate power gain and voltage gain. The square of the magnitude of the incident and reflected waves represent power. Therefore, the power delivered to the load in Fig. 2.6.5 is given by the difference between the incident and reflected power, namely

$$P_L = |b_2|^2 - |a_2|^2 = |b_2|^2(1 - |\Gamma_L|^2) \qquad (2.7.3)$$

The power available from a source is defined as the power delivered by the source to a conjugately matched load. Figure 2.7.3 shows the signal flow graph of a source connected to a conjugate match load (i.e., $\Gamma_L = \Gamma_s^*$). Therefore, the power available from the source, in Fig. 2.7.3, is given by

$$P_{AVS} = |b_g|^2 - |a_g|^2 \qquad (2.7.4)$$

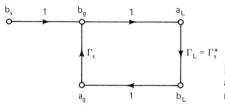

**Figure 2.7.3** Signal flow graph of a voltage source connected to a conjugate matched load.

Observing that $b_g = b_s + b_g \Gamma_s \Gamma_s^*$ and $a_g = b_g \Gamma_s^*$, we obtain

$$b_g = \frac{b_s}{1 - |\Gamma_s|^2} \tag{2.7.5}$$

and

$$a_g = \frac{b_s \Gamma_s^*}{1 - |\Gamma_s|^2} \tag{2.7.6}$$

Substituting (2.7.5) and (2.7.6) into (2.7.4) gives

$$P_{\text{AVS}} = \frac{|b_s|^2}{1 - |\Gamma_s|^2} \tag{2.7.7}$$

The previous results could have also been obtained as follows. Observe that the power delivered to the load $\Gamma_L$ in Fig. 2.7.3 is given by

$$P_L = |a_L|^2 (1 - |\Gamma_L|^2) = \frac{|b_s|^2 (1 - |\Gamma_L|^2)}{|1 - \Gamma_s \Gamma_L|^2}$$

Therefore, with $\Gamma_L = \Gamma_s^*$ the power delivered to the load is equal to the available power from the source, and (2.7.7) follows.

The transducer power gain, called $G_T$, is defined as the ratio of the power delivered to a load to the power available from the source. From (2.7.3) and (2.7.7) we obtain

$$G_T = \frac{P_L}{P_{\text{AVS}}} = \frac{|b_2|^2}{|b_s|^2}(1 - |\Gamma_L|^2)(1 - |\Gamma_s|^2) \tag{2.7.8}$$

The ratio $b_2/b_s$ can be obtained using Mason's rule, namely

$$\frac{b_2}{b_s} = \frac{S_{21}}{1 - (S_{11}\Gamma_s + S_{22}\Gamma_L + S_{21}\Gamma_L S_{12}\Gamma_s) + S_{11}\Gamma_s S_{22}\Gamma_L} \tag{2.7.9}$$

Substituting (2.7.9) into (2.7.8) results in

$$G_T = \frac{|S_{21}|^2(1 - |\Gamma_s|^2)(1 - |\Gamma_L|^2)}{|(1 - S_{11}\Gamma_s)(1 - S_{22}\Gamma_L) - S_{21}S_{12}\Gamma_L\Gamma_s|^2} \tag{2.7.10}$$

The denominator of (2.7.10) can be further manipulated and $G_T$ can be expressed in the form

$$G_T = \frac{1 - |\Gamma_s|^2}{|1 - \Gamma_{\text{IN}}\Gamma_s|^2} |S_{21}|^2 \frac{1 - |\Gamma_L|^2}{|1 - S_{22}\Gamma_L|^2} \tag{2.7.11}$$

or

$$G_T = \frac{1-|\Gamma_s|^2}{|1-S_{11}\Gamma_s|^2}|S_{21}|^2 \frac{1-|\Gamma_L|^2}{|1-\Gamma_{OUT}\Gamma_L|^2} \qquad (2.7.12)$$

where $\Gamma_{IN}$ and $\Gamma_{OUT}$ are given by (2.7.1) and (2.7.2), respectively.

The voltage gain of the amplifier is defined as the ratio of the output voltage to the input voltage. That is,

$$A_v = \frac{a_2 + b_2}{a_1 + b_1}$$

and dividing by $b_s$ gives

$$A_v = \frac{a_2/b_s + b_2/b_s}{a_1/b_s + b_1/b_s}$$

Therefore, we need to calculate the ratios $a_2/b_s$, $b_2/b_s$, $a_1/b_s$, and $b_1/b_s$ using Mason's rule. The expression for $A_v$ can be shown to be

$$A_v = \frac{S_{21}(1+\Gamma_L)}{(1-S_{22}\Gamma_L) + S_{11}(1-S_{22}\Gamma_L) + S_{21}\Gamma_L S_{12}} \qquad (2.7.13)$$

## PROBLEMS

2.1. (a) Show that impedances having a negative real part (i.e., $z = -r + jx$) have a reflection coefficient whose magnitude is greater than 1.
   (b) Prove that negative resistances can be handled in the Smith chart by plotting $1/\Gamma^*$ and interpreting the resistance circles as being negative and the reactance circles as marked.
   (c) Locate in the Smith chart the impedances $Z_1 = -20 + j16$ Ω and $Z_2 = -200 + j25$ Ω and find the associated reflection coefficient. Normalize the impedances to 50 Ω.
   (d) Work the problem in part (c) in the compressed Smith chart.

2.2. (a) Prove that the maximum normalized resistance in a transmission line is numerically equal to the VSWR.
   (b) Prove that the minimum normalized resistance in a transmission line is numerically equal to 1/VSWR.

2.3 Show that the impedance along a transmission line repeats itself at every $\lambda/2$ distance. That is,

$$Z(d) = Z\left(d + \frac{n\lambda}{2}\right), \qquad n = 1, 2, 3, \ldots$$

2.4. Show that the impedance along a transmission line can be expressed in the form

$$Z(d) = R(d) + jX(d) = |Z(d)|e^{j\theta_d}$$

where

$$R(d) = Z_o \frac{1 - |\Gamma|^2}{1 - 2|\Gamma|\cos\phi + |\Gamma|^2}$$

$$X(d) = Z_o \frac{2|\Gamma|\sin\phi}{1 - 2|\Gamma|\cos\phi + |\Gamma|^2}$$

$$|Z(d)| = Z_0 \sqrt{\frac{1 + 2|\Gamma|\cos\psi + |\Gamma|^2}{1 - 2|\Gamma|\cos\phi + |\Gamma|^2}}$$

$$\theta_d = \tan^{-1}\frac{X(d)}{R(d)} = \tan^{-1}\left(\frac{2|\Gamma|\sin\phi}{1 - |\Gamma|^2}\right)$$

$$\Gamma = |\Gamma_0|e^{j\phi}, \quad \Gamma_o = |\Gamma_o|e^{j\phi_l}, \quad \phi = \phi_l - 2\beta d$$

**2.5.** Find the input impedance, the load reflection coefficient, and the VSWR in a transmission line having an electrical length of 90°, $Z_o = 50\ \Omega$, and terminated in the load $Z_L = 50 + j100\ \Omega$. Work the problem in both the Z and Y Smith charts.

**2.6.** (a) Design a single-stub matching system (see Fig. P2.6) to match the load $Z_L = 15 + j25\ \Omega$ to a 50-Ω transmission line. The characteristic impedance of the short-circuited stub is 50 Ω.

(b) Design the single-stub matching system in Fig. P2.6 assuming that the characteristic impedance of the stub is 100 Ω.

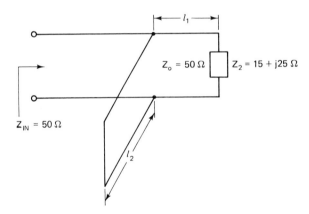

**Figure P2.6**

**2.7.** Two types of ell matching networks are shown in Fig. P2.7. Select one that can match the load $Y_L = (8 - j12) \times 10^{-3}$ S to a 50-Ω transmission line. Find the element values at $f = 1$ GHz.

**2.8.** Design four different ell matching networks to match the load $Z_L = 10 + j40\ \Omega$ to a 50-Ω transmission line.

**2.9.** Design the matching network in Fig. P2.9 that provides $Y_L = (4 - j4) \times 10^{-3}$ S to the transistor. Find the element values at 700 MHz.

**2.10.** Use (2.5.8) to (2.5.11) to show that for Duroid, ($\varepsilon_r = 2.23$, $h = 0.7874$ mm), a 50-Ω characteristic impedance is obtained with $W/h = 3.073$. Also, $\varepsilon_{ff} = 1.91$ and $\lambda = 0.7236\lambda_0$.

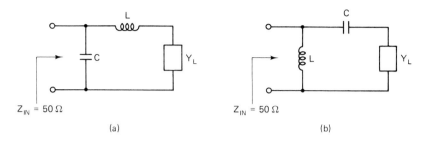

**Figure P2.7**

**2.11.** In the amplifier shown in Fig. 2.5.8b, calculate the width and the length of the lines at $f = 1$ GHz.

**2.12.** Design two microstrip matching networks for an amplifier whose reflection coefficients at $f = 800$ MHz, in a 50-$\Omega$ system, are $\Gamma_s = 0.8\lfloor 160°$ and $\Gamma_L = 0.7\lfloor 20°$. Show the diagram for the complete amplifier using balanced shunt stubs.

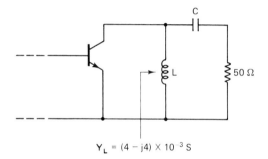

**Figure P2.9**

**2.13.** Design the matching networks in Fig. P2.13 to match the load $Z_L = 100 + j100$ $\Omega$ to a 50-$\Omega$ transmission line.

**2.14.** Design the matching network in Fig. P2.14 to match a 50-$\Omega$ load to the impedance $Z_{IN} = 25 - j25$ $\Omega$.

**2.15.** Verify the expressions for $G_T$ in (2.7.11) and (2.7.12).

**2.16.** Verify the expression for $A_v$ in (2.7.13).

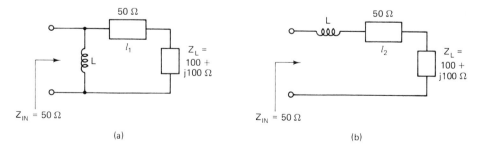

**Figure P2.13**

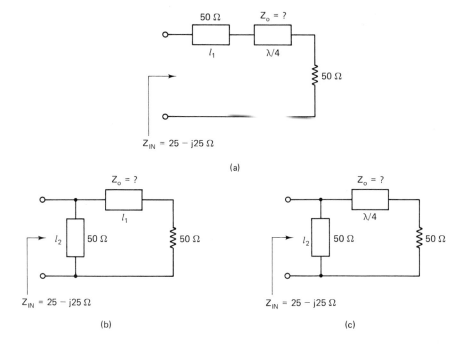

**Figure P2.14**

## REFERENCES

[2.1] I. J. Bahl and D. K. Trivedi, "A Designer's Guide to Microstrip Line," *Microwaves*, May 1977.

[2.2] H. Sobol, "Applications of Integrated Circuit Technology to Microwave Frequencies," *Proceedings of the IEEE*, August 1971.

[2.3] H. Sobol, "Extending IC Technology to Microwave Equipment," *Electronics*, March 1967.

# 3

# MICROWAVE TRANSISTOR AMPLIFIER DESIGN

## 3.1 INTRODUCTION

This chapter develops some basic principles used in the analysis and design of microwave transistor amplifiers. Based on the $S$ parameters of the transistor and certain performance requirements, a systematic procedure is developed for the design of a microwave transistor amplifier. Of course, some errors are expected in the final implementation of the design resulting from parameter variations, stray capacitances, and other random causes.

The most important design considerations in a microwave transistor amplifier are stability, power gain, bandwidth, noise, and dc requirements. This chapter deals with the problems of stability, power gain, and the dc bias design.

A design usually starts with a set of specifications and the selection of the proper transistor. Then a systematic mathematical solution, aided by graphical methods, is developed to determine the transistor loading (i.e., the source and load reflection coefficients) for a particular stability and gain criteria. An unconditionally stable transistor will not oscillate with any passive termination. On the other hand, a design using a potentially unstable transistor requires some analysis and careful considerations so that the passive terminations produce a stable amplifier.

Design procedures for both unilateral and bilateral transistors, based on stability and gain requirements, are described. Both passive and active dc bias networks for BJTs and GaAs FETs are analyzed. It is important to select the correct dc operating point and the proper dc network topology in order to obtain the desired ac performance.

## 3.2 POWER GAIN EQUATIONS

Several power gain equations appear in the literature and are used in the design of microwave amplifiers. Figure 3.2.1 illustrates a microwave amplifier signal flow graph and the different powers used in gain equations are indicated. The transducer power gain $G_T$, the power gain $G_p$ (also called the operating power gain), and the available power gain $G_A$ are defined as follows:

$$G_T = \frac{P_L}{P_{AVS}} = \frac{\text{power delivered to the load}}{\text{power available from the source}}$$

$$G_p = \frac{P_L}{P_{IN}} = \frac{\text{power delivered to the load}}{\text{power input to the network}}$$

and

$$G_A = \frac{P_{AVN}}{P_{AVS}} = \frac{\text{power available from the network}}{\text{power available from the source}}$$

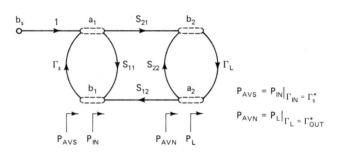

**Figure 3.2.1** Different power definitions.

The expressions for $G_T$ were already derived in (2.7.11) and (2.7.12). Observing that $P_{IN} = |a_1|^2 - |b_1|^2$ and that the power available from the network, $P_{AVN}$, is that power delivered by the network to a conjugately matched load, we can write the power gain equations in the form

$$G_T = \frac{1 - |\Gamma_s|^2}{|1 - \Gamma_{IN}\Gamma_s|^2} |S_{21}|^2 \frac{1 - |\Gamma_L|^2}{|1 - S_{22}\Gamma_L|^2} \qquad (3.2.1)$$

$$G_T = \frac{1 - |\Gamma_s|^2}{|1 - S_{11}\Gamma_s|^2} |S_{21}|^2 \frac{1 - |\Gamma_L|^2}{|1 - \Gamma_{OUT}\Gamma_L|^2} \qquad (3.2.2)$$

$$G_p = \frac{1}{1 - |\Gamma_{IN}|^2} |S_{21}|^2 \frac{1 - |\Gamma_L|^2}{|1 - S_{22}\Gamma_L|^2} \qquad (3.2.3)$$

$$G_A = \frac{1 - |\Gamma_s|^2}{|1 - S_{11}\Gamma_s|^2} |S_{21}|^2 \frac{1}{1 - |\Gamma_{OUT}|^2} \qquad (3.2.4)$$

## Sec. 3.2  Power Gain Equations

$$\Gamma_{IN} = S_{11} + \frac{S_{12} S_{21} \Gamma_L}{1 - S_{22} \Gamma_L} \quad (3.2.5)$$

$$\Gamma_{OUT} = S_{22} + \frac{S_{12} S_{21} \Gamma_s}{1 - S_{11} \Gamma_s} \quad (3.2.6)$$

$G_T$ is a $f(\Gamma_s, \Gamma_L, [S])$ (i.e., a function of $\Gamma_s$, $\Gamma_L$, and the $S$ parameters of the transistor), $G_p = f(\Gamma_L, [S])$, and $G_A = f(\Gamma_s, [S])$.

If we assume the network to be unilateral, that is, when $S_{12} = 0$, then $\Gamma_{IN} = S_{11}$, $\Gamma_{OUT} = S_{22}$, and the unilateral transducer power gain from (3.2.1) and (3.2.2), called $G_{TU}$, is given by

$$G_{TU} = \frac{1 - |\Gamma_s|^2}{|1 - S_{11} \Gamma_s|^2} |S_{21}|^2 \frac{1 - |\Gamma_L|^2}{|1 - S_{22} \Gamma_L|^2} \quad (3.2.7)$$

The first term in (3.2.7) depends on the $S_{11}$ parameter of the transistor and the source reflection coefficient. The second term, $|S_{21}|^2$, depends on the transistor scattering parameter $S_{21}$; and the third term depends on the $S_{22}$ parameter of the transistor and the load reflection coefficient. We can think of (3.2.7) as being composed of three distinct and independent gain terms. Therefore, we can write (3.2.7) in the form

$$G_{TU} = G_s G_o G_L \quad (3.2.8)$$

where

$$G_s = \frac{1 - |\Gamma_s|^2}{|1 - S_{11} \Gamma_s|^2} \quad (3.2.9)$$

$$G_o = |S_{21}|^2 \quad (3.2.10)$$

$$G_L = \frac{1 - |\Gamma_L|^2}{|1 - S_{22} \Gamma_L|^2} \quad (3.2.11)$$

and the microwave amplifier can be represented by three different gain (or loss) blocks, as shown in Fig. 3.2.2.

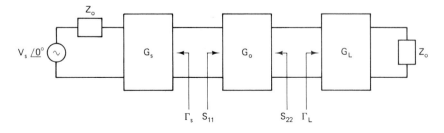

**Figure 3.2.2**  Unilateral transducer power gain block diagram.

The terms $G_s$ and $G_L$ represent the gain or loss produced by the matching or mismatching of the input or output circuits, respectively. The term $G_s$ affects the degree of matching or mismatching between $\Gamma_s$ and $S_{11}$. Although the $G_s$ block is made up of passive components, it can either have a gain contribution greater than unity or a loss. The reason we usually refer to $G_s$ as a gain block is that there is an intrinsic mismatch loss between $Z_o$, the matching network, and $S_{11}$ (i.e., between $\Gamma_s$ and $S_{11}$). Therefore, decreasing the mismatch loss can be thought of as providing a gain. Similarly, the term $G_L$ affects the output matching and can be thought of as the output gain block. The term $G_o$ is related to the device and is equal to $|S_{21}|^2$. In terms of decibels we can write from (3.2.8) to (3.2.11)

$$G_{TU} \text{ (dB)} = G_s \text{ (dB)} + G_o \text{ (dB)} + G_L \text{ (dB)}$$

If we optimize $\Gamma_s$ and $\Gamma_L$ to provide maximum gain in $G_s$ and $G_L$, we refer to the gain as the maximum unilateral transducer power gain, called $G_{TU,\max}$. The maximum gain of $G_S$ and $G_L$, with $|S_{11}| < 1$ and $|S_{22}| < 1$, is obtained when

$$\Gamma_s = S_{11}^*$$

and

$$\Gamma_L = S_{22}^*$$

Therefore, from (3.2.9) and (3.2.11) we obtain

$$G_{s,\max} = \frac{1}{1 - |S_{11}|^2}$$

$$G_{L,\max} = \frac{1}{1 - |S_{22}|^2}$$

and (3.2.8) gives

$$G_{TU,\max} = G_{s,\max} G_o G_{L,\max}$$
$$= \frac{1}{1 - |S_{11}|^2} |S_{21}|^2 \frac{1}{1 - |S_{22}|^2} \quad (3.2.12)$$

The appropriate block diagram for (3.2.12) is shown in Fig. 3.2.3.

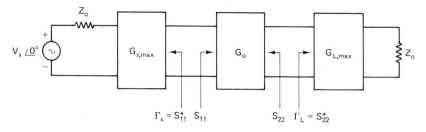

**Figure 3.2.3** Maximum unilateral transducer power gain block diagram.

## 3.3 STABILITY CONSIDERATIONS

The stability of an amplifier, or its resistance to oscillate, is a very important consideration in a design and can be determined from the $S$ parameters, the matching networks, and the terminations. In a two-port network, oscillations are possible when either the input or output port presents a negative resistance. This occurs when $|\Gamma_{IN}| > 1$ or $|\Gamma_{OUT}| > 1$, which for a unilateral device occurs when $|S_{11}| > 1$ or $|S_{22}| > 1$.

The two-port network shown in Fig. 3.3.1 is said to be unconditionally stable at a given frequency if the real parts of $Z_{IN}$ and $Z_{OUT}$ are greater than zero for all passive load and source impedances. If the two-port is not unconditionally stable, it is potentially unstable. That is, some passive load and source terminations can produce input and output impedances having a negative real part.

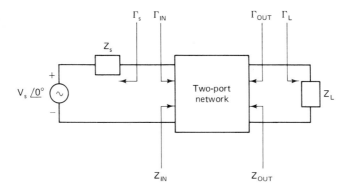

**Figure 3.3.1** Stability of two-port networks.

In terms of reflection coefficients, the conditions for unconditional stability at a given frequency are

$$|\Gamma_s| < 1 \tag{3.3.1}$$

$$|\Gamma_L| < 1 \tag{3.3.2}$$

$$|\Gamma_{IN}| = \left| S_{11} + \frac{S_{12}S_{21}\Gamma_L}{1 - S_{22}\Gamma_L} \right| < 1 \tag{3.3.3}$$

and

$$|\Gamma_{OUT}| = \left| S_{22} + \frac{S_{12}S_{21}\Gamma_s}{1 - S_{11}\Gamma_s} \right| < 1 \tag{3.3.4}$$

where, of course, all coefficients are normalized to the same characteristic impedance $Z_o$.

The solutions of (3.3.1) to (3.3.4) give the required conditions for the two-port network to be unconditionally stable. However, before we discuss the intricacies of the necessary and sufficient conditions for unconditional stability, a graphical analysis of (3.3.1) to (3.3.4) is presented. The graphical analysis is especially useful in the analysis of potentially unstable transistors.

When the two-port in Fig. 3.3.1 is potentially unstable, there may be values of $\Gamma_s$ and $\Gamma_L$ (i.e., source and load impedances) for which the real parts of $Z_{IN}$ and $Z_{OUT}$ are positive. These values of $\Gamma_s$ and $\Gamma_L$ (i.e., regions in the Smith chart) can be determined using the following graphical procedure.

First, the regions where values of $\Gamma_L$ and $\Gamma_s$ produce $|\Gamma_{IN}| = 1$ and $|\Gamma_{OUT}| = 1$ are determined, respectively. Setting the magnitude of (3.3.3) and (3.3.4) equal to 1 and solving for the values of $\Gamma_L$ and $\Gamma_s$ shows that the solutions for $\Gamma_L$ and $\Gamma_s$ lie on circles (called *stability circles*) whose equations are given by

$$\left| \Gamma_L - \frac{(S_{22} - \Delta S_{11}^*)^*}{|S_{22}|^2 - |\Delta|^2} \right| = \left| \frac{S_{12} S_{21}}{|S_{22}|^2 - |\Delta|^2} \right| \qquad (3.3.5)$$

and

$$\left| \Gamma_s - \frac{(S_{11} - \Delta S_{22}^*)^*}{|S_{11}|^2 - |\Delta|^2} \right| = \left| \frac{S_{12} S_{21}}{|S_{11}|^2 - |\Delta|^2} \right| \qquad (3.3.6)$$

where

$$\Delta = S_{11} S_{22} - S_{12} S_{21}$$

The radii and centers of the circles where $|\Gamma_{IN}| = 1$ and $|\Gamma_{OUT}| = 1$ in the $\Gamma_L$ plane and $\Gamma_s$ plane, respectively, are obtained from (3.3.5) and (3.3.6), namely:

$\Gamma_L$ values for $|\Gamma_{IN}| = 1$ (*Output Stability Circle*):

$$r_L = \left| \frac{S_{12} S_{21}}{|S_{22}|^2 - |\Delta|^2} \right| \qquad \text{(radius)} \qquad (3.3.7)$$

$$C_L = \frac{(S_{22} - \Delta S_{11}^*)^*}{|S_{22}|^2 - |\Delta|^2} \qquad \text{(center)} \qquad (3.3.8)$$

$\Gamma_s$ values for $|\Gamma_{OUT}| = 1$ (*Input Stability Circle*):

$$r_s = \left| \frac{S_{12} S_{21}}{|S_{11}|^2 - |\Delta|^2} \right| \qquad \text{(radius)} \qquad (3.3.9)$$

$$C_s = \frac{(S_{11} - \Delta S_{22}^*)^*}{|S_{11}|^2 - |\Delta|^2} \qquad \text{(center)} \qquad (3.3.10)$$

With the $S$ parameters of a two-port device at one frequency, the expressions (3.3.7) to (3.3.10) can be calculated, plotted on a Smith chart, and the set of values of $\Gamma_L$ and $\Gamma_s$ that produce $|\Gamma_{IN}| = 1$ and $|\Gamma_{OUT}| = 1$ easily observed. Figure 3.3.2 illustrates the graphical construction of the stability circles

Sec. 3.3  Stability Considerations

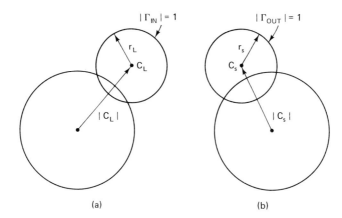

**Figure 3.3.2** Stability circle construction in the Smith chart: (a) $\Gamma_L$ plane; (b) $\Gamma_s$ plane.

where $|\Gamma_{IN}| = 1$ and $|\Gamma_{OUT}| = 1$. On one side of the stability circle boundary, in the $\Gamma_L$ plane, we will have $|\Gamma_{IN}| < 1$ and on the other side $|\Gamma_{IN}| > 1$. Similarly, in the $\Gamma_s$ plane on one side of the stability circle boundary, we will have $|\Gamma_{OUT}| < 1$ and on the other side $|\Gamma_{OUT}| > 1$.

Next, we need to determine which area in the Smith chart represents the stable region. In other words, the region where values of $\Gamma_L$ (where $|\Gamma_L| < 1$) produce $|\Gamma_{IN}| < 1$, and where values of $\Gamma_s$ (where $|\Gamma_s| < 1$) produce $|\Gamma_{OUT}| < 1$. To this end, we observe that if $Z_L = Z_o$, then $\Gamma_L = 0$ and from (3.2.5) $|\Gamma_{IN}| = |S_{11}|$. If the magnitude of $S_{11}$ is less than 1, then $|\Gamma_{IN}| < 1$ when $\Gamma_L = 0$. That is, the center of the Smith chart in Fig. 3.3.2a represents a stable operating point, because for $\Gamma_L = 0$ it follows that $|\Gamma_{IN}| < 1$. On the other hand, if $|S_{11}| > 1$ when $Z_L = Z_o$, then $|\Gamma_{IN}| > 1$ when $\Gamma_L = 0$ and the center of the Smith chart represents an unstable operating point. Figure 3.3.3 illus-

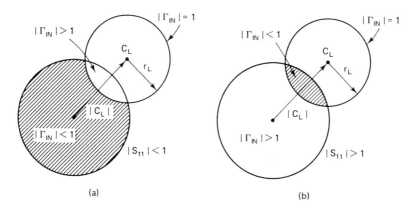

**Figure 3.3.3** Smith chart illustrating stable and unstable regions in the $\Gamma_L$ plane.

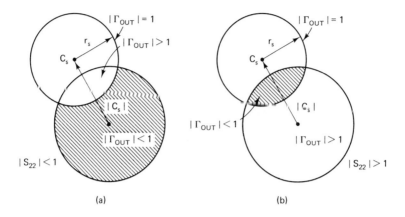

**Figure 3.3.4** Smith chart illustrating stable and unstable regions in the $\Gamma_s$ plane.

trates the two cases discussed. The shaded area represents the values of $\Gamma_L$ that produce a stable operation. Similarly, Fig. 3.3.4 illustrates stable and unstable regions for $\Gamma_s$.

For unconditional stability any passive load or source in the network must produce a stable condition. From a graphical point of view, for $|S_{11}| < 1$ and $|S_{22}| < 1$, we want the stability circles shown in Figs. 3.3.3a and 3.3.4a to fall completely outside (or to completely enclose) the Smith chart. The case where the stability circles fall completely outside the Smith chart is illustrated in Fig. 3.3.5. Therefore, the conditions for unconditional stability for all passive sources and loads can be expressed in the form

$$||C_L| - r_L| > 1 \quad \text{for } |S_{11}| < 1 \quad (3.3.11)$$

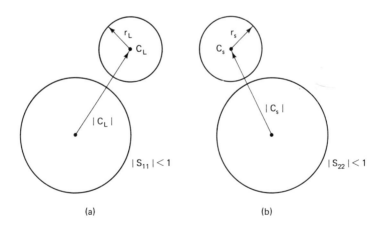

**Figure 3.3.5** Conditions for unconditional stability: (a) $\Gamma_L$ plane; (b) $\Gamma_s$ plane.

Sec. 3.3   Stability Considerations

and

$$\|C_s| - r_s| > 1 \quad \text{for } |S_{22}| < 1 \tag{3.3.12}$$

If either $|S_{11}| > 1$ or $|S_{22}| > 1$, the network cannot be unconditionally stable because the termination $\Gamma_L = 0$ or $\Gamma_s = 0$ [see (3.3.3) and (3.3.4)] will produce $|\Gamma_{IN}| > 1$ or $|\Gamma_{OUT}| > 1$.

We now return to the necessary and sufficient conditions for a two-port to be unconditionally stable. A straightforward but somewhat lengthy manipulation of (3.3.1) to (3.3.4) results in the following necessary and sufficient conditions for unconditional stability [3.1]:

$$K = \frac{1 - |S_{11}|^2 - |S_{22}|^2 + |\Delta|^2}{2|S_{12}S_{21}|} > 1 \tag{3.3.13}$$

and

$$1 - |S_{11}|^2 > |S_{12}S_{21}| \tag{3.3.14}$$

$$1 - |S_{22}|^2 > |S_{12}S_{21}| \tag{3.3.15}$$

Alternatively, in Ref. [3.2] the necessary and sufficient conditions for unconditional stability are given as

$$K > 1$$

and

$$B_1 = 1 + |S_{11}|^2 - |S_{22}|^2 - |\Delta|^2 > 0 \tag{3.3.16}$$

It can be shown (see Problem 3.5) that both sets of conditions are equivalent and that a simpler set of conditions is [3.3]

$$K > 1$$

and

$$|\Delta| < 1 \tag{3.3.17}$$

where

$$|\Delta| = |S_{11}S_{22} - S_{12}S_{21}|$$

A derivation of the conditions (3.3.13) to (3.3.15) and also of $K > 1$ and $|\Delta| < 1$, starting from (3.3.1) to (3.3.4) and using the bilinear transformation, is given in Ref. [3.4].

In conclusion, when $|S_{11}| < 1$ and $|S_{22}| < 1$, the two-port network is unconditionally stable if $K > 1$ and $|\Delta| < 1$, where $K$ is given by (3.3.13) and $|\Delta|$ by (3.3.17). The condition $|\Delta| < 1$ can be shown to be similar to (3.3.14) and (3.3.15), or to (3.3.16) (see Problem 3.5).

For example, a transistor whose $S$ parameters are $S_{11} = 0.75 \underline{|-60°}$, $S_{21} = 6 \underline{|90°}$, $S_{22} = 0.5 \underline{|60°}$, and $S_{12} = 0.3 \underline{|70°}$ has $K = 1.344$ and $|\Delta| = 2.156$. This transistor has $K > 1$; however, it is potentially unstable because $|\Delta| > 1$.

In fact, the input and output stability circles, from (3.3.7) to (3.3.10), are located at $C_s = 0.1 \underline{|107.4°}$, $r_s = 0.44$, $C_L = 0.26 \underline{|-36.3°}$, and $r_L = 0.41$, which can be drawn inside the Smith chart to show the unstable regions. Also, a calculation of $B_1$ from (3.3.16) produces a negative value (i.e., $B_1 = -3.34$). Since $K > 1$ and $B_1 < 0$, we conclude again that the transistor is potentially unstable.

In the unilateral amplifier $K \to \infty$ and we have unconditional stability if $|S_{11}| < 1$ and $|S_{22}| < 1$ for all passive load and source terminations.

In the potentially unstable situation illustrated in Figs. 3.3.3 and 3.3.4, the real part of the input and output impedances can be negative for some source and load reflection coefficients. In this case, selecting $\Gamma_s$ and $\Gamma_L$ in the stable region produces a stable operation.

Even when the selection of $\Gamma_L$ and $\Gamma_s$ produces $|\Gamma_{IN}| > 1$ or $|\Gamma_{OUT}| > 1$, the circuit can be made stable if the total input and output loop resistance in Fig. 3.3.1 is positive. In other words, the circuit is stable if

$$\text{Re}\,(Z_s + Z_{IN}) > 0$$

and

$$\text{Re}\,(Z_L + Z_{OUT}) > 0$$

A potentially unstable transistor can be made unconditionally stable by either resistively loading the transistor or by adding negative feedback. These techniques are not recommended in narrowband amplifiers because of the resulting degradation in power gain, noise figure, and so on. Narrowband amplifier design with potentially unstable transistors is best done by the proper selection of $\Gamma_s$ and $\Gamma_L$ to ensure stability. On the other hand, the techniques are popular in the design of some broadband amplifiers where the transistor is potentially unstable.

The following example illustrates how resistive loading can stabilize a potentially unstable transistor.

**Example 3.3.1**

The $S$ parameters of a transistor at $f = 800$ MHz are

$$S_{11} = 0.65 \underline{|-95°}$$
$$S_{12} = 0.035 \underline{|40°}$$
$$S_{21} = 5 \underline{|115°}$$
$$S_{22} = 0.8 \underline{|-35°}$$

Determine the stability and show how resistive loading can stabilize the transistor.

**Solution.** From (3.3.13) and (3.3.17) we find that $K = 0.547$ and $\Delta = 0.504 \underline{|249.6°}$. Since $K < 1$, the transistor is potentially unstable at $f = 800$ MHz.

The input and output stability circles are calculated using (3.3.7) to (3.3.10):

$$C_s = 1.79 \underline{|122°} \qquad C_L = 1.3 \underline{|48°}$$
$$r_s = 1.04 \qquad r_L = 0.45$$

Sec. 3.3  Stability Considerations

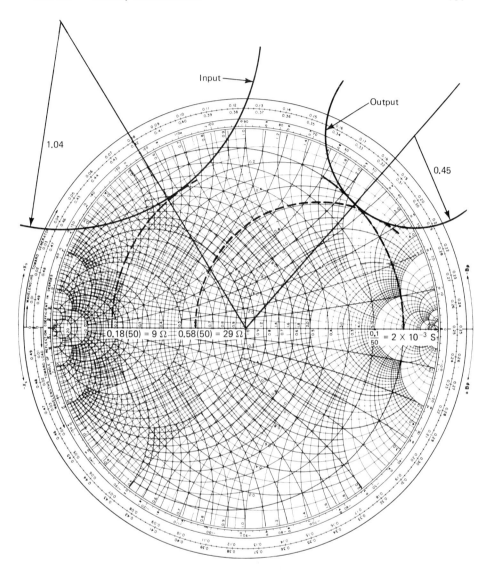

**Figure 3.3.6** Input and output stability circles.

Figure 3.3.6 shows the plot of the stability circles, together with the stable region. It can be seen that a series resistor with the input, of approximately 9 Ω, assures stability at the input. Also, a series resistor of approximately 29 Ω, or a shunt resistor of approximately 500 Ω at the output, produces stability at the output. The three choices of resistive loading are shown in Fig. 3.3.7. The most popular is the shunt resistor configuration in Fig. 3.3.7c.

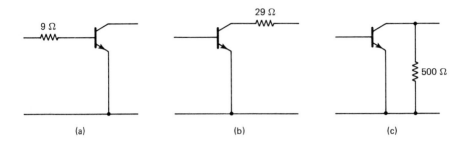

**Figure 3.3.7** Three types of resistive loading to improve stability.

For the stabilized shunt resistor configuration in Fig. 3.3.7c (i.e., with a 500-Ω shunt resistor), the resulting $S$ parameters are

$$S_{11} = 0.65 \underline{/-94°}$$
$$S_{12} = 0.032 \underline{/41.2°}$$
$$S_{21} = 4.62 \underline{/116.2°}$$
$$S_{22} = 0.66 \underline{/-36°}$$

and from (3.3.13) and (3.3.17) $K = 1.04$ and $\Delta = 0.409 \underline{/250.13°}$, which show that the stabilized network in Fig. 3.3.7c is unconditionally stable at $f = 800$ MHz.

Negative feedback can be used to stabilize a transistor by neutralizing $S_{12}$, that is, by making $S_{12} = 0$. However, this is not commonly done. In a broadband amplifier design using a potentially unstable transistor, a common procedure is to use resistive loading to stabilize the transistor and negative feedback to provide the proper ac performance, that is, to provide constant gain and low input and output VSWR.

## 3.4 CONSTANT-GAIN CIRCLES—UNILATERAL CASE

The unilateral transducer power gain is given by (3.2.7) or (3.2.8), and the maximum unilateral transducer power gain, obtained when $\Gamma_s = S_{11}^*$ and $\Gamma_L = S_{22}^*$, is given by (3.2.12). The expressions for $G_s$ and $G_L$ in (3.2.9) and (3.2.11) are similar in form and they can be written in the general form

$$G_i = \frac{1 - |\Gamma_i|^2}{|1 - S_{ii}\Gamma_i|^2} \qquad (3.4.1)$$

where $i = s$ ($ii = 11$) or $i = L$ ($ii = 22$). The design for a specific gain is based on (3.4.1).

Two cases must be considered in the analysis of (3.4.1), the unconditionally stable case where $|S_{ii}| < 1$ and the potentially unstable case where $|S_{ii}| > 1$.

Sec. 3.4   Constant-Gain Circles—Unilateral Case   103

**Unconditionally Stable Case, $|S_{ii}| < 1$**

The maximum value of (3.4.1) is obtained when $\Gamma_i = S_{ii}^*$, and it is given by

$$G_{i,\max} = \frac{1}{1 - |S_{ii}|^2} \qquad (3.4.2)$$

The terminations that produce $G_{i,\max}$ are called the *optimum terminations*.

From (3.4.1), $G_i$ has a minimum value of zero when $|\Gamma_i| = 1$. Other values of $\Gamma_i$ produce values of $G_i$ between zero and $G_{i,\max}$, that is,

$$0 \leq G_i \leq G_{i,\max}$$

The values of $\Gamma_i$ that produce a constant gain $G_i$ will be shown to lie in a circle in the Smith chart. These circles are called *constant-gain circles*.

Define the normalized gain factor as

$$g_i = \frac{G_i}{G_{i,\max}} = G_i(1 - |S_{ii}|^2) \qquad (3.4.3)$$

such that

$$0 \leq g_i \leq 1$$

In order to set $g_i$ equal to a constant and solve for the values of $\Gamma_i$, we let

$$\Gamma_i = U_i + jV_i \qquad (3.4.4)$$

and

$$S_{ii} = A_{ii} + jB_{ii} \qquad (3.4.5)$$

Substitute (3.4.4) and (3.4.5) into (3.4.3); after some manipulations we obtain

$$\left[ U_i - \frac{g_i A_{ii}}{1 - |S_{ii}|^2(1 - g_i)} \right]^2 + \left[ V_i + \frac{g_i B_{ii}}{1 - |S_{ii}|^2(1 - g_i)} \right]^2 = \left[ \frac{\sqrt{1 - g_i}\,(1 - |S_{ii}|^2)}{1 - |S_{ii}|^2(1 - g_i)} \right]^2 \qquad (3.4.6)$$

Equation (3.4.6) is recognized as a family of circles with $g_i$ as a parameter. The centers of the circles are located at

$$U_c = \frac{g_i A_{ii}}{1 - |S_{ii}|^2(1 - g_i)}$$

and

$$V_c = \frac{-g_i B_{ii}}{1 - |S_{ii}|^2(1 - g_i)}$$

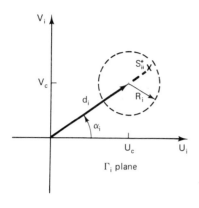

**Figure 3.4.1** Constant-gain circles in the Smith chart.

and the radii of the circles are given by

$$R_i = \frac{\sqrt{1-g_i}\,(1-|S_{ii}|^2)}{1-|S_{ii}|^2(1-g_i)} \qquad (3.4.7)$$

Figure 3.4.1 illustrates the location of a constant-gain circle. The distance from the origin to the center of a constant-gain circle is given by

$$d_i = \sqrt{U_c^2 + V_c^2} = \frac{g_i\,|S_{ii}|}{1-|S_{ii}|^2(1-g_i)} \qquad (3.4.8)$$

and the angle of inclination, $\alpha_i$, is

$$\tan \alpha_i = \frac{V_c}{U_c}$$

or

$$\alpha_i = \tan^{-1} \frac{-B_{ii}}{A_{ii}} \qquad (3.4.9)$$

Equation (3.4.9) shows that the constant-gain circles are located, at a distance $d_i$ given by (3.4.8), along the line drawn from the origin to the point $S_{ii}^*$ (i.e., $S_{ii}^* = A_{ii} - jB_{ii}$).

It is observed that when $g_i = 1$ (i.e., when $G_i = G_{i,\max}$) (3.4.7) gives $R_i = 0$ and (3.4.8) gives $d_i = |S_{ii}|$. Therefore, the constant-gain circle for maximum gain is represented by a point, located at $S_{ii}^*$.

In conclusion, the procedure for drawing the constant-gain circles in the Z Smith chart is:

1. Locate $S_{ii}^*$ and draw a line from the origin to $S_{ii}^*$. At $S_{ii}^*$, the gain is $G_{i,\max}$ and given by (3.4.2).
2. Determine the values of $G_i$, where $0 \le G_i \le G_{i,\max}$, for which the constant-gain circles are to be drawn, and calculate the corresponding values of $g_i = G_i/G_{i,\max}$.

Sec. 3.4  Constant-Gain Circles—Unilateral Case

3. From (3.4.8), determine the values of $d_i$ for each $g_i$.
4. From (3.4.7), determine the values of $R_i$ for each $g_i$.

The 0-dB circle ($G_i = 1$) always passes through the origin of the Smith chart. This is not a coincidence. In fact, $G_i = 1$ occurs when $\Gamma_i = 0$ and from (3.4.3)

$$g_{i,0dB} = 1 - |S_{ii}|^2$$

then, from (3.4.7) and (3.4.8),

$$R_{i,0dB} = d_{i,0dB} = \frac{|S_{ii}|}{1 + |S_{ii}|^2}$$

which shows that the radius and the distance from the origin to the center of the 0-dB constant-gain circle are identical.

A typical set of constant-gain circles for $G_s$ are calculated in the following example and shown in Fig. 3.4.2.

**Example 3.4.1**

The S parameters of a BJT measured at $V_{CE} = 10$ V, $I_C = 30$ mA, and $f = 1$ GHz, in a 50-Ω system are

$$S_{11} = 0.73 \underline{|175°}$$
$$S_{12} = 0$$
$$S_{21} = 4.45 \underline{|65°}$$
$$S_{22} = 0.21 \underline{|-80°}$$

(a) Calculate the optimum terminations.
(b) Calculate $G_{s,max}$, $G_{L,max}$, and $G_{TU,max}$ in decibels.
(c) Draw several $G_s$ constant-gain circles.
(d) Design the input matching network for $G_s = 2$ dB.

**Solution.** (a) The optimum terminations are

$$\Gamma_s = S_{11}^* = 0.73 \underline{|-175°}$$

and

$$\Gamma_L = S_{22}^* = 0.21 \underline{|80°}$$

Using the Smith chart, the impedances associated with $\Gamma_s$ and $\Gamma_L$ are $Z_s = 50(0.152 - j0.047) = 7.6 - j2.35$ Ω and $Z_L = 50(0.97 + j0.43) = 48.5 + j21.5$ Ω.

(b) From (3.4.2) we find that

$$G_{s,max} = \frac{1}{1 - |S_{11}|^2} = 2.141 \quad \text{or} \quad 3.31 \text{ dB}$$

$$G_{L,max} = \frac{1}{1 - |S_{22}|^2} = 1.046 \quad \text{or} \quad 0.196 \text{ dB}$$

Since
$$G_o = |S_{21}|^2 = 19.8 \quad \text{or} \quad 12.97 \text{ dB}$$
then
$$G_{TU,\max} \text{ (dB)} = 3.31 + 0.196 + 12.97 = 16.47 \text{ dB}$$

(c) The output network provides little gain (i.e., $G_{L,\max} = 0.195$ dB); therefore, the output matching network is designed to present the optimum termination $\Gamma_L = 0.21\,\underline{|80°}$. Since $G_{s,\max} = 3.31$ dB, constant-gain circles at 2, 1, 0, and $-1$ dB are drawn in Fig. 3.4.2a. The necessary calculations are given in Fig. 3.4.2b.

(d) Any $\Gamma_s$ along the $G_s = 2$ dB circle provides the constant gain. Selecting $\Gamma_s$ at point $A$ (i.e., $z_s = 0.42 + j0.1$) in Fig. 3.4.2a results in the input matching network shown in Fig. 3.4.3a. The details of the matching network design are shown in Fig. 3.4.3b.

### Potentially Unstable Case, $|S_{ii}| > 1$

In this case $|S_{ii}| > 1$ and it is possible for a passive termination to produce an infinite value of $G_i$. The infinite value of $G_i$ in (3.4.1) is produced by the critical value of $\Gamma_i$, called $\Gamma_{i,c}$, given by

$$\Gamma_{i,c} = \frac{1}{S_{ii}} \qquad (3.4.10)$$

Equation (3.4.10) basically states that the real part of the impedance associated with $\Gamma_{i,c}$ is equal to the magnitude of the negative resistance associated with $S_{ii}$. Therefore, the total input or output loop resistance is zero, and oscillations will occur.

As discussed in Section 2.2, in the case of negative resistances we can locate $1/S_{ii}^*$ in the Smith chart and interpret the resistance circles as being negative and the reactance circles as labeled.

With $g_i$ defined as in (3.4.3), namely

$$g_i = G_i(1 - |S_{ii}|^2)$$

where now $g_i$ can attain negative values because $|S_{ii}| > 1$, it is not difficult to show that the centers of the constant-gain circles are located at the distance $d_i$ given by (3.4.8) and the radii of the circles $R_i$ given by (3.4.7). The centers of the constant-gain circles are located, at the distance $d_i$, along a line drawn from the origin to the point $1/S_{ii}$.

To prevent oscillations in the input or output port, $\Gamma_i$ must be selected such that the real part of the termination impedance is larger than the magnitude of the negative resistance associated with the point $1/S_{ii}^*$. The stable region is that region where values of $\Gamma_i$ produce a termination such that

$$\text{Re}(Z_s) > |\text{Re}(Z_{IN})|$$

and

$$\text{Re}(Z_L) > |\text{Re}(Z_{OUT})|$$

## Sec. 3.4 Constant-Gain Circles—Unilateral Case

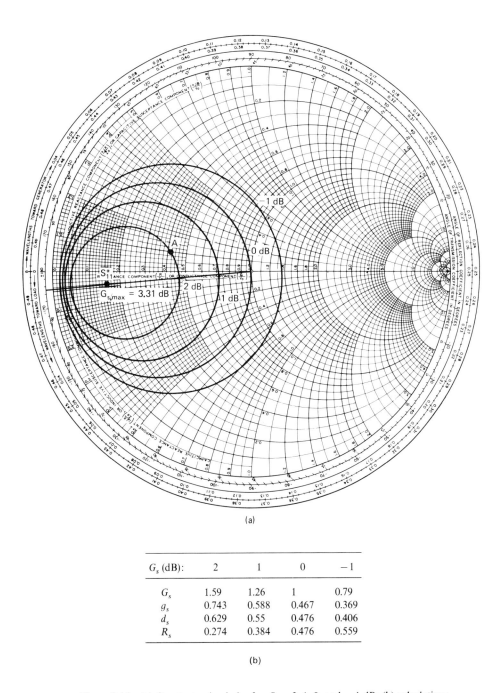

(a)

| $G_s$ (dB): | 2 | 1 | 0 | −1 |
|---|---|---|---|---|
| $G_s$ | 1.59 | 1.26 | 1 | 0.79 |
| $g_s$ | 0.743 | 0.588 | 0.467 | 0.369 |
| $d_s$ | 0.629 | 0.55 | 0.476 | 0.406 |
| $R_s$ | 0.274 | 0.384 | 0.476 | 0.559 |

(b)

**Figure 3.4.2** (a) Constant-gain circles for $G_s = 2, 1, 0,$ and $-1$ dB; (b) calculations of constant-gain circles.

**108** Microwave Transistor Amplifier Design  Chap. 3

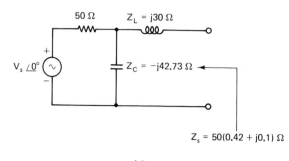

(a)

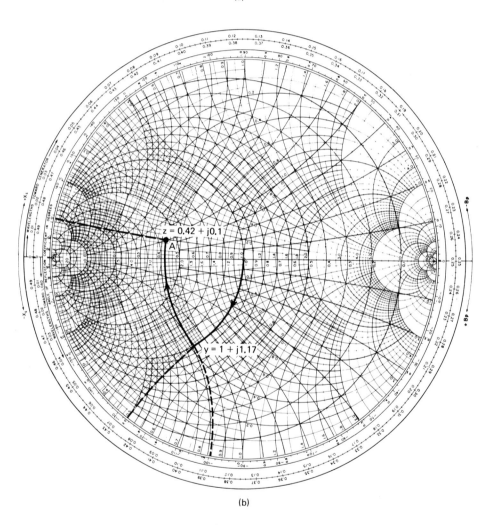

(b)

**Figure 3.4.3**  Input matching network for $G_s = 2$ dB.

## Sec. 3.4 Constant-Gain Circles—Unilateral Case

A typical construction is illustrated in Fig. 3.4.4, where the critical value of $\Gamma_s$ (i.e., $\Gamma_{s,c} = 1/S_{11}$) and two constant-gain circles are shown.

**Example 3.4.2**

The S parameters of a GaAs FET measured at $V_{DS} = 5$ V, $I_{DS} = 10$ mA, and $f = 1$ GHz in a 50-$\Omega$ system are

$$S_{11} = 2.27 \underline{|-120°}$$

$$S_{12} = 0$$

$$S_{21} = 4 \underline{|50°}$$

$$S_{22} = 0.6 \underline{|-80°}$$

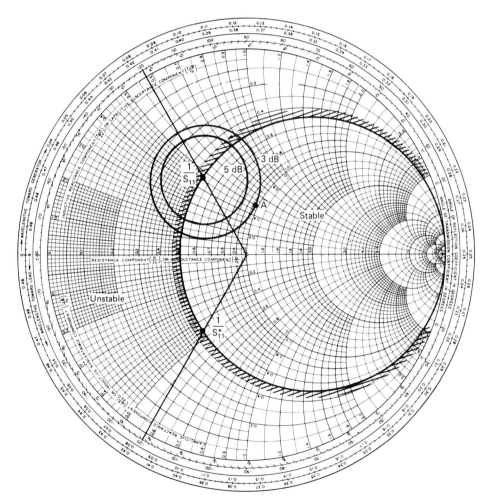

**Figure 3.4.4** Stable and unstable regions, and $G_s$ constant-gain circles for Example 3.4.2.

(a) Calculate the input impedance and the optimum output terminations.
(b) Determine the unstable region in the Smith chart and construct constant-gain circles for $G_s = 5$ dB and $G_s = 3$ dB.
(c) Design the input matching network for $G_s = 3$ dB with the greatest degree of stability.
(d) Determine $G_{TU}$ in decibels.

**Solution.** (a) The input impedance is obtained from the Smith chart at the point $1/S_{11}^* = 0.44 \underline{|-120°}$ (see Fig. 3.4.4), namely

$$Z_{IN} = 50(-0.5 - j0.46) = -25 - j23 \; \Omega$$

The optimum termination for $G_L$ is

$$\Gamma_L = S_{22}^* = 0.6 \underline{|80°}$$

The impedance associated with $\Gamma_L$ is obtained from the Smith chart as $Z_L = 50(0.56 + j1.03) = 28 + j51.5 \; \Omega$.

(b) The unstable region is where Re $(Z_s) < |\text{Re} (Z_{IN})|$. The unstable region is marked in Fig. 3.4.4.

In order to construct the constant-gain circle for $G_s = 5$ dB, we first locate the point $1/S_{11}$ in Fig. 3.4.4. Then, from (3.4.3), (3.4.7), and (3.4.8), we find that

$$g_s = 3.16[1 - (2.27)^2] = -13.123$$

$$R_s = \frac{\sqrt{1 + 13.123} \; [1 - (2.27)^2]}{1 - (2.27)^2(1 + 13.123)} = 0.217$$

and

$$d_s = \frac{-13.123(2.27)}{1 - (2.27)^2(1 + 13.123)} = 0.415$$

The $G_s = 5$ dB circle is drawn in Fig. 3.4.4. Similarly, for the $G_s = 3$ dB circle, we find that $g_s = -8.286$, $d_s = 0.401$, and $R_s = 0.27$.

(c) In order to obtain the greatest degree of stability we select $\Gamma_s$ on the $G_s = 3$ dB circle such that it has the largest positive real part. That is, $\Gamma_s$ is selected at point $A$ in Fig. 3.4.4, namely

$$\Gamma_s = 0.245 \underline{|79°}$$

or

$$Z_s = 50(0.97 + j0.5) = 48.5 + j25 \; \Omega$$

Since the input loop resistance is $48.5 - 25 = 23.5 \; \Omega$, the input port is stable.

(d) Since $G_s = 3$ dB,

$$G_{L,\text{max}} = \frac{1}{1 - |S_{22}|^2} = \frac{1}{1 - (0.6)^2} = 1.562 \quad \text{or} \quad 1.94 \text{ dB}$$

and

$$G_o = |S_{21}|^2 = (4)^2 = 16 \quad \text{or} \quad 12.04 \text{ dB}$$

the unilateral transducer gain is

$$G_{TU} \text{ (dB)} = 3 + 12.04 + 1.94 = 16.98 \text{ dB}$$

## 3.5 UNILATERAL FIGURE OF MERIT

When $S_{12}$ can be set equal to zero, the design procedure is much simpler. In order to determine the error involved in assuming $S_{12} = 0$, we form the magnitude ratio of $G_T$ and $G_{TU}$ from (2.7.10) and (3.2.7), namely

$$\frac{G_T}{G_{TU}} = \frac{1}{|1-X|^2} \quad (3.5.1)$$

where

$$X = \frac{S_{12} S_{21} \Gamma_s \Gamma_L}{(1 - S_{11}\Gamma_s)(1 - S_{22}\Gamma_L)}$$

From (3.5.1) the ratio of the transducer power gain to the unilateral transducer power gain is bounded by

$$\frac{1}{(1+|X|)^2} < \frac{G_T}{G_{TU}} < \frac{1}{(1-|X|)^2}$$

When $\Gamma_s = S_{11}^*$ and $\Gamma_L = S_{22}^*$, $G_{TU}$ has a maximum value and, in this case, the maximum error introduced when using $G_{TU}$ is bounded by

$$\frac{1}{(1+U)^2} < \frac{G_T}{G_{TU}} < \frac{1}{(1-U)^2} \quad (3.5.2)$$

where

$$U = \frac{|S_{12}||S_{21}||S_{11}||S_{22}|}{(1-|S_{11}|^2)(1-|S_{22}|^2)} \quad (3.5.3)$$

is known as the *unilateral figure of merit*.

The value of $U$ varies with frequency because of its dependence on the $S$ parameters. A typical variation of $U$ with frequency is shown in Fig. 3.5.1. In this case, the maximum value of $U$ occurs at 100 MHz and 1 GHz, and is given by $U = -15$ dB or $U = 0.03$. Therefore, from (3.5.2),

$$\frac{1}{(1+0.03)^2} < \frac{G_T}{G_{TU}} < \frac{1}{(1-0.03)^2}$$

or in decibels,

$$-0.26 \text{ dB} < \frac{G_T}{G_{TU}} < 0.26 \text{ dB}$$

and the maximum error is $\pm 0.26$ dB at 100 MHz and 1 GHz. In this case, the error is small enough to justify the unilateral assumption.

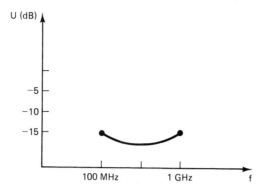

**Figure 3.5.1** Frequency dependence of the unilateral figure of merit.

## 3.6 SIMULTANEOUS CONJUGATE MATCH— BILATERAL CASE

When $S_{12} \neq 0$, and the unilateral assumption cannot be made, the input and output reflection coefficients are given by (3.2.5) and (3.2.6), respectively. The conditions required to obtain maximum transducer power gain are

$$\Gamma_{IN} = \Gamma_s^* \tag{3.6.1}$$

and

$$\Gamma_{OUT} = \Gamma_L^* \tag{3.6.2}$$

These conditions are illustrated in Fig. 3.6.1.

From (3.2.5), (3.2.6), (3.6.1), and (3.6.2) we can write

$$\Gamma_s^* = S_{11} + \frac{S_{12} S_{21} \Gamma_L}{1 - S_{22} \Gamma_L} \tag{3.6.3}$$

and

$$\Gamma_L^* = S_{22} + \frac{S_{12} S_{21} \Gamma_s}{1 - S_{11} \Gamma_s} \tag{3.6.4}$$

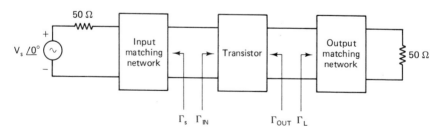

**Figure 3.6.1** Simultaneous conjugate match exists when $\Gamma_{IN} = \Gamma_s^*$ and $\Gamma_{OUT} = \Gamma_L^*$.

## Sec. 3.6  Simultaneous Conjugate Match—Bilateral Case

Solving (3.6.3) and (3.6.4) simultaneously gives the values of $\Gamma_s$ and $\Gamma_L$ required for a simultaneous conjugate match. Calling these values $\Gamma_{Ms}$ and $\Gamma_{ML}$, we obtain

$$\Gamma_{Ms} = \frac{B_1 \pm \sqrt{B_1^2 - 4|C_1|^2}}{2C_1} \quad (3.6.5)$$

$$\Gamma_{ML} = \frac{B_2 \pm \sqrt{B_2^2 - 4|C_2|^2}}{2C_2} \quad (3.6.6)$$

$$B_1 = 1 + |S_{11}|^2 - |S_{22}|^2 - |\Delta|^2 \quad (3.6.7)$$

$$B_2 = 1 + |S_{22}|^2 - |S_{11}|^2 - |\Delta|^2 \quad (3.6.8)$$

$$C_1 = S_{11} - \Delta S_{22}^*$$

$$C_2 = S_{22} - \Delta S_{11}^*$$

In what follows we will show that for an unconditionally stable two-port network, the solutions with a minus sign in (3.6.5) and (3.6.6) are the useful ones.

If $|B_1/2C_1| > 1$ and $B_1 > 0$ in (3.6.5), the solution with the minus sign produces $|\Gamma_{Ms}| < 1$ and the solution with the plus sign produces $|\Gamma_{Ms}| > 1$. If $|B_1/2C_1| > 1$ with $B_1 < 0$ in (3.6.5), the solution with the plus sign produces $|\Gamma_{Ms}| < 1$ and the solution with the minus sign produces $|\Gamma_{Ms}| > 1$. Similar considerations apply to (3.6.6).

Since it can be shown that $|B_i/2C_i|^2 > 1$ ($i = 1$ or 2) is similar to $K^2 > 1$, it follows that the condition $|B_i/2C_i| > 1$ is similar to $|K| > 1$ (see Problem 3.13).

Therefore, if $|K| > 1$ with $K$ positive, one solution of (3.6.5) and (3.6.6) has a magnitude less than 1, and the other solution has magnitude greater than 1. In fact, in this case if $B_i > 0$ the solutions with the minus sign have magnitudes less than 1 (observe from Problem 3.5 that in this case $B_1$ and $B_2$ have the same sign). The analysis for $|K| > 1$ with $K$ negative is left as an exercise (see Problem 3.13).

Associated with $\Gamma_{Ms}$ and $\Gamma_{ML}$ are a source and a load impedance. The real parts of these impedances are positive if $|\Gamma_{Ms}| < 1$ and $|\Gamma_{ML}| < 1$. From the previous considerations, we conclude that in terms of $K$, the condition that a two-port network can be simultaneously matched with $|\Gamma_{Ms}| < 1$ and $|\Gamma_{ML}| < 1$ is [3.1, 3.2]

$$K > 1$$

The condition $K > 1$ is only a necessary condition for unconditional stability. Therefore, a simultaneous conjugate match having unconditional stability is possible if $K > 1$ and $|\Delta| < 1$. Since $|\Delta| < 1$ implies that $B_1 > 0$ and $B_2 > 0$, the minus signs must be used in (3.6.5) and (3.6.6) when calculating the simultaneous conjugate match for an unconditionally stable two-port network.

In what follows any reference to a simultaneous conjugate match assumes that the two-port network is unconditionally stable. In a potentially unstable situation the design procedure is best done in terms of $G_p$ or $G_A$ (see Section 3.8).

The maximum transducer power gain, under simultaneous conjugate match conditions, is obtained from (2.7.10) with $\Gamma_s = \Gamma_{Ms}$ and $\Gamma_L = \Gamma_{ML}$, namely

$$G_{T,\max} = \frac{(1 - |\Gamma_{Ms}|^2)|S_{21}|^2(1 - |\Gamma_{ML}|^2)}{|(1 - S_{11}\Gamma_{Ms})(1 - S_{22}\Gamma_{ML}) - S_{12}S_{21}\Gamma_{ML}\Gamma_{Ms}|^2} \qquad (3.6.9)$$

Substituting (3.6.5) and (3.6.6) into (3.6.9), and using (3.3.13) gives the relation

$$G_{T,\max} = \frac{|S_{21}|}{|S_{12}|}(K - \sqrt{K^2 - 1}) \qquad (3.6.10)$$

The maximum stable gain is defined as the value of $G_{T,\max}$ when $K = 1$, namely

$$G_{\text{MSG}} = \frac{|S_{21}|}{|S_{12}|} \qquad (3.6.11)$$

$G_{\text{MSG}}$ is a figure of merit that represents the maximum value that $G_{T,\max}$ can have. It can be achieved by resistively loading the two-port (i.e., the transistor) to make $K = 1$, or by using feedback.

## 3.7 CONSTANT-GAIN CIRCLES—BILATERAL CASE

The bilateral case occurs when $S_{12}$ cannot be neglected. Two different cases must be considered in the analysis: the unconditionally stable bilateral case and the potentially unstable bilateral case.

### Unconditionally Stable Bilateral Case, $K > 1$ and $|\Delta| < 1$

This situation occurs when $K > 1$ and $|\Delta| < 1$, and any passive source and load terminations can be used. Of course, the terminations $\Gamma_{Ms}$ and $\Gamma_{ML}$, given by (3.6.5) and (3.6.6), will produce a simultaneous conjugate match which results in the maximum value of the transducer power gain.

If the design calls for a transducer power gain different from the maximum, a constant-gain circle procedure based on (3.2.1) or (3.2.2) can be used. For example, a procedure based on (3.2.1) is as follows. Write (3.2.1) in the form

$$G_T = G'_s G_o G_L \qquad (3.7.1)$$

where

$$G'_s = \frac{1 - |\Gamma_s|^2}{|1 - \Gamma_{IN}\Gamma_s|^2} \qquad (3.7.2)$$

$$G_o = |S_{21}|^2$$

and

$$G_L = \frac{1 - |\Gamma_L|^2}{|1 - S_{22}\Gamma_L|^2} \qquad (3.7.3)$$

Then the design procedure is as follows:

1. From (3.7.3), the constant-gain circles for $G_L$ can be drawn using (3.4.3) and (3.4.7) to (3.4.9). Select the desired $\Gamma_L$ for a given $G_L$ gain.
2. Calculate $\Gamma_{IN}$ from (3.2.5). Observe that $\Gamma_{IN}$ depends on $\Gamma_L$; therefore, $G'_s$ depends on $G_L$.
3. From (3.7.2), the constant-gain circles for $G'_s$ can be drawn using (3.4.3) and (3.4.7) to (3.4.9) (observing that $\Gamma_{IN}$ replaces $S_{ii}$). Select the desired $\Gamma_s$ for a given $G'_s$ gain. The value of $G'_s$ might not be satisfactory and will require the selection of another $\Gamma_L$ and the procedure repeated.
4. Design the matching networks.

The procedure just outlined is not recommended for a practical design since $\Gamma_{IN}$ is a function of $\Gamma_L$, making the $G'_s$ function dependent of the $G_L$ function. Furthermore, the centers of the gain circles do not give $G_{T,\max}$. In fact, the graphical approach becomes tedious because of the iterative process required for obtaining the desired gain.

As shown in the next section, the design of a microwave transistor amplifier in the unconditional stable bilateral case, for a gain different from $G_{T,\max}$, can be done using the operating power gain equation.

### Example 3.7.1

Design a microwave amplifier using a GaAs FET to operate at $f = 6$ GHz with maximum transducer power gain. The transistor $S$ parameters at the linear bias point, $V_{DS} = 4$ V and $I_{DS} = 0.5 I_{DSS}$, are

$$S_{11} = 0.641 \underline{|-171.3°}$$
$$S_{12} = 0.057 \underline{|16.3°}$$
$$S_{21} = 2.058 \underline{|28.5°}$$
$$S_{22} = 0.572 \underline{|-95.7°}$$

**Solution.** From (3.3.13) and (3.3.17) we obtain $K = 1.504$ and $\Delta = 0.3014 \underline{|109.88°}$. Since $K > 1$ and $|\Delta| < 1$, the GaAs FET is unconditionally stable.

Next, we must decide if the amplifier can be considered unilateral. From (3.5.3), $U = 0.1085$, and from (3.5.2),

$$-0.89 \text{ dB} < \frac{G_T}{G_{TU}} < 1 \text{ dB}$$

The inequality above shows that $S_{12}$ cannot be neglected.

The reflection coefficients for a simultaneous conjugate match are calculated from (3.6.5) and (3.6.6) as follows:

$$B_1 = 0.9928$$
$$B_2 = 0.8255$$
$$C_1 = 0.4786 \underline{|182.7°}$$
$$C_2 = 0.3911 \underline{|256.1°}$$
$$\Gamma_{Ms} = 0.762 \underline{|177.3°}$$

and

$$\Gamma_{ML} = 0.718 \underline{|103.9°}$$

The maximum transducer power gain, from (3.6.10), is

$$G_{T,\max} = \frac{2.058}{0.057}(1.504 - \sqrt{(1.504)^2 - 1}) = 13.74 \quad \text{or} \quad 11.38 \text{ dB}$$

The design of the matching networks using microstrip lines is illustrated in Fig. 3.7.1, where the admittances associated with $\Gamma_{Ms}$ and $\Gamma_{ML}$ are

$$Y_{Ms} = \frac{7.2 - j1.23}{50} = (144 - j24.6) \times 10^{-3} \text{ S}$$

and

$$Y_{ML} = \frac{0.414 - j1.19}{50} = (8.28 - j23.8) \times 10^{-3} \text{ S}$$

The input matching network can be designed with an open shunt stub of length $0.185\lambda$ and a series transmission line of length $0.0615\lambda$. The output matching network is designed with an open shunt stub of length $0.176\lambda$ and a series transmission line of length $0.169\lambda$.

The ac amplifier schematic is shown in Fig. 3.7.2. Using Duroid ($\varepsilon_r = 2.23$, $h = 0.7874$ mm) for the board material, we find that $W = 2.41$ mm for a characteristic impedance of 50 Ω, $\varepsilon_{ff} = 1.9052$, and $\lambda = 0.7245\lambda_0$, where $\lambda_0 = 5$ cm at $f = 6$ GHz. The microstrip lengths at $f = 6$ GHz are

$$0.185\lambda = 6.70 \text{ mm}$$

$$0.0615\lambda = 2.23 \text{ mm}$$

Sec. 3.7  Constant-Gain Circles—Bilateral Case                                      117

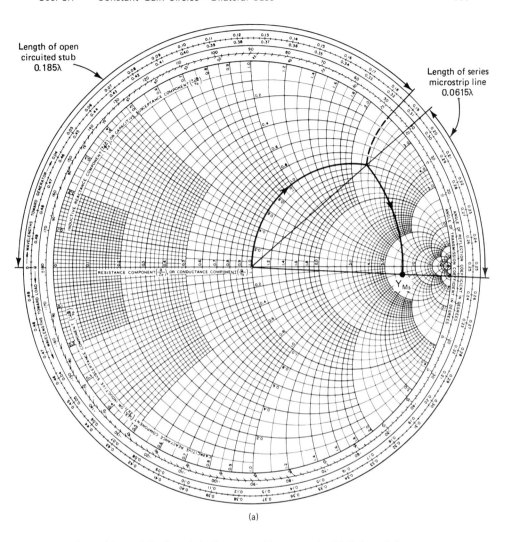

(a)

**Figure 3.7.1** (a) Design of the input matching network; (b) design of the output matching network.

$$0.169\lambda = 6.12 \text{ mm}$$

$$0.176\lambda = 6.38 \text{ mm}$$

The design for $G_{T,\text{max}}$ with $\Gamma_{Ms}$ and $\Gamma_{ML}$, at 6 GHz, assures that the input and output VSWR are 1.

This example is revisited in Section 4.3, where noise considerations are included. Finally, we should point out that the stability must be checked at all frequencies, so that the reflection coefficients $\Gamma_{Ms}$ and $\Gamma_{ML}$ provide stable operation.

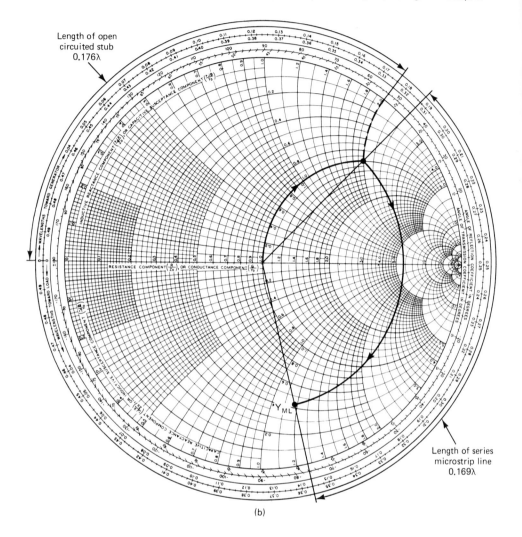

**Figure 3.7.1** (*continued*)

## Potentially Unstable Bilateral Case, $K < 1$ or $|\Delta| > 1$

A design procedure based on the expression for $G_T$ in (3.2.1) or (3.2.2) is not recommended because it leads to a tedious iterative process. The process is somewhat similar to that given for the unconditionally stable bilateral case [i.e., (3.7.1) to (3.7.3)].

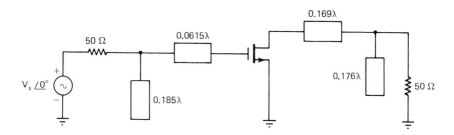

**Figure 3.7.2** The ac schematic of a GaAs FET microwave amplifier. All microstrip lines have a characteristic impedance of 50 Ω.

As shown in the next section, a design procedure for this case can be developed using the operating power gain equation.

In conclusion, when designing a microwave transistor amplifier where $S_{12}$ cannot be neglected (i.e., bilateral case), a design procedure that calls for a transducer power gain different from $G_{T,\max}$ is tedious, and a simpler design procedure can be developed using the operating power gain equation. When the transistor is unconditionally stable, a simultaneous conjugate match can be found, and the design procedure is based on $G_{T,\max}$ or (as shown in the next section) on the operating power gain. In fact, in this case both procedures lead to the same results.

## 3.8 OPERATING AND AVAILABLE POWER GAIN CIRCLES

When $S_{12}$ cannot be neglected, a design procedure based on the operating power gain $G_p$ is commonly used. The operating power gain is independent of the source impedance; therefore, an operating power gain circle procedure for both unconditionally stable and potentially unstable transistors is simple and recommended for practical designs.

Again we must consider two cases, the unconditionally stable case and the potentially unstable case.

### Unconditionally Stable Bilateral Case

To develop a design procedure with $G_p$, we write (3.2.3) in the form

$$G_p = \frac{|S_{21}|^2(1 - |\Gamma_L|^2)}{\left(1 - \left|\frac{S_{11} - \Delta\Gamma_L}{1 - S_{22}\Gamma_L}\right|^2\right)|1 - S_{22}\Gamma_L|^2}$$

$$= |S_{21}|^2 g_p \qquad (3.8.1)$$

where

$$g_p = \frac{1 - |\Gamma_L|^2}{|1 - S_{22}\Gamma_L|^2 - |S_{11} - \Delta\Gamma_L|^2}$$

$$= \frac{1 - |\Gamma_L|^2}{1 - |S_{11}|^2 + |\Gamma_L|^2(|S_{22}|^2 - |\Delta|^2) - 2\operatorname{Re}(\Gamma_L C_2)} \quad (3.8.2)$$

and

$$C_2 = S_{22} - \Delta S_{11}^* \quad (3.8.3)$$

Here $G_p$ and $g_p$ are functions of the device S parameters and $\Gamma_L$.

The constant operating power gain circles are obtained by letting $\Gamma_L = U_L + jV_L$ and substituting into (3.8.2). After some manipulations we obtain

$$\left[U_L - \frac{g_p \operatorname{Re}[C_2^*]}{1 + g_p(|S_{22}|^2 - |\Delta|^2)}\right]^2 + \left[V_L - \frac{g_p \operatorname{Im}[C_2^*]}{1 + g_p(|S_{22}|^2 - |\Delta|^2)}\right]^2$$

$$= \left[\frac{[1 - 2K|S_{12}S_{21}|g_p + |S_{12}S_{21}|^2 g_p^2]^{1/2}}{1 + g_p(|S_{22}|^2 - |\Delta|^2)}\right]^2 \quad (3.8.4)$$

Equation (3.8.4) is recognized as a family of circles in the $U_L:V_L$ plane (i.e., the Smith chart) with $g_p$ as a parameter. The centers of the circles are located at

$$U_p = \frac{g_p \operatorname{Re}[C_2^*]}{1 + g_p(|S_{22}|^2 - |\Delta|^2)}$$

and

$$V_p = \frac{g_p \operatorname{Im}[C_2^*]}{1 + g_p(|S_{22}|^2 - |\Delta|^2)}$$

The radii of the circles are given by

$$R_p = \frac{[1 - 2K|S_{12}S_{21}|g_p + |S_{12}S_{21}|^2 g_p^2]^{1/2}}{|1 + g_p(|S_{22}|^2 - |\Delta|^2)|} \quad (3.8.5)$$

The distance from the origin of the Smith chart to the centers of the circles is given by

$$d_p = \sqrt{U_p^2 + V_p^2} = \frac{g_p|C_2^*|}{|1 + g_p(|S_{22}|^2 - |\Delta|^2)|}$$

Therefore, the centers of the circles $C_p$ can be written as

$$C_p = \frac{g_p C_2^*}{1 + g_p(|S_{22}|^2 - |\Delta|^2)} \quad (3.8.6)$$

The maximum operating power gain occurs when $R_p = 0$. Therefore,

Sec. 3.8  Operating and Available Power Gain Circles

from (3.8.5) we can write

$$g_{p,\max}^2 |S_{12} S_{21}|^2 - 2K|S_{12} S_{21}| g_{p,\max} + 1 = 0 \qquad (3.8.7)$$

where $g_{p,\max}$ is the maximum value of $g_p$. The solution to (3.8.7) for unconditional stability is

$$g_{p,\max} = \frac{1}{|S_{12} S_{21}|} (K - \sqrt{K^2 - 1}) \qquad (3.8.8)$$

Therefore, substituting (3.8.8) into (3.8.1) gives

$$G_{p,\max} = \frac{|S_{21}|}{|S_{12}|} (K - \sqrt{K^2 - 1})$$

For a given $G_p$, $\Gamma_L$ is selected from the constant operating power gain circles. $G_{p,\max}$ results when $\Gamma_L$ is selected at the distance where $g_{p,\max} = G_{p,\max}/|S_{21}|^2$. The maximum output power results when a conjugate match is selected at the input (i.e., $\Gamma_s = \Gamma_{IN}^*$). It also follows that when $\Gamma_s = \Gamma_{IN}^*$ the input power is equal to the maximum available input power. Therefore, under these circumstances the maximum transducer power gain ($G_{T,\max}$) and the operating power gain are equal, and the values of $\Gamma_s$ and $\Gamma_L$ that result in $G_{p,\max}$ are identical to $\Gamma_{Ms}$ and $\Gamma_{ML}$, respectively.

The procedure for drawing a constant operating power gain circle in the Z Smith chart is as follows:

1. For a given $G_p$, the radius and center of the constant operating power gain circle are given by (3.8.5) and (3.8.6).
2. Select the desired $\Gamma_L$.
3. For the given $\Gamma_L$, maximum output power is obtained with a conjugate match at the input, namely with $\Gamma_s = \Gamma_{IN}^*$, where $\Gamma_{IN}$ is given by (3.2.5). This value of $\Gamma_s$ produces the transducer power gain $G_T = G_p$.

**Example 3.8.1**

Design the amplifier in Example 3.7.1 to have an operating power gain of 9 dB instead of $G_{T,\max} = G_{p,\max} = 11.38$ dB.

**Solution.** Since

$$|S_{21}|^2 = (2.058)^2 = 4.235 \quad \text{or} \quad 6.27 \text{ dB}$$

then

$$g_p = \frac{G_p}{|S_{21}|^2} = \frac{7.94}{4.235} = 1.875$$

From the results in Example 3.7.1, $K = 1.504$, $|\Delta| = 0.3014$, and $C_2 = 0.3911 \underline{/256.1°}$. Therefore, the radius and center of the 9-dB operating power gain circle, from (3.8.5) and (3.8.6), are $R_p = 0.431$ and $C_p = 0.508 \underline{/103.9°}$.

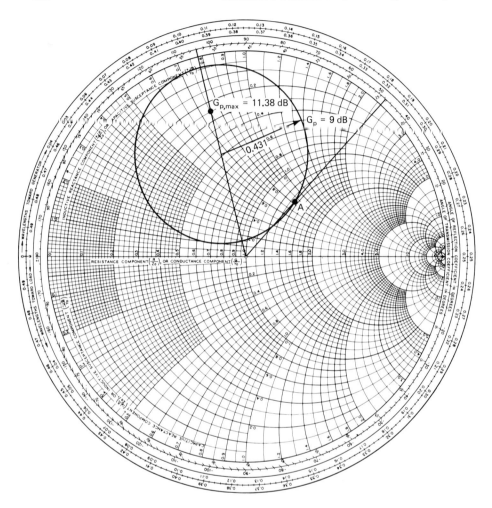

**Figure 3.8.1** Operating power gain circle for $G_p = 9$ dB and location of $G_{p,\max} = 11.38$ dB.

The graphical construction is shown in Fig. 3.8.1. The 9-dB operating power gain circle shows all loads that produce $G_p = 9$ dB. The load reflection coefficient can be selected at point $A$, namely $\Gamma_L = 0.36\,\underline{/47.5°}$. Then the required $\Gamma_s$ for maximum output power is

$$\Gamma_s = \Gamma_{IN}^* = \left[S_{11} + \frac{S_{12}S_{21}\Gamma_L}{1 - S_{22}\Gamma_L}\right]^* = 0.629\,\underline{/175.51°}$$

The location of $G_{p,\max} = 11.38$ dB can be found as follows:

$$g_{p,\max} = \frac{G_{p,\max}}{|S_{21}|^2} = \frac{13.74}{(2.058)^2} = 3.24$$

$$R_{p,\max} = 0$$

and

$$C_{p,max} = \frac{g_{p,max} C_2^*}{1 + g_{p,max}(|S_{22}|^2 - |\Delta|^2)} = 0.718 \underline{/103.9°}$$

At the location of $G_{p,max}$, we obtain from Fig. 3.8.1 $\Gamma_{L,max} = 0.718 \underline{/103.9°}$. This value of $\Gamma_{L,max}$ is identical to the value $\Gamma_{ML}$ found in Example 3.7.1. The associated $\Gamma_s$ for maximum output power is

$$\Gamma_{s,max} = \left[S_{11} + \frac{S_{12} S_{21} \Gamma_{L,max}}{1 - S_{22} \Gamma_{L,max}}\right]^* = 0.762 \underline{/177.3°}$$

which is identical to the value of $\Gamma_{Ms}$ in Example 3.7.1.

The derivation of the constant available power gain circles is similar to that of the operating power gain circles. It is simple to show that in the $\Gamma_s$ plane the radius $R_a$ and center $C_a$ of a circle can be calculated using the relations

$$g_a = \frac{G_A}{|S_{21}|^2} \tag{3.8.9}$$

$$C_1 = S_{11} - \Delta S_{22}^* \tag{3.8.10}$$

$$R_a = \frac{[1 - 2K|S_{12} S_{21}| g_a + |S_{12} S_{21}|^2 g_a^2]^{1/2}}{|1 + g_a(|S_{11}|^2 - |\Delta|^2)|} \tag{3.8.11}$$

and

$$C_a = \frac{g_a C_1^*}{1 + g_a(|S_{11}|^2 - |\Delta|^2)} \tag{3.8.12}$$

For a given $G_A$, a constant available power gain circle can be plotted using (3.8.9) to (3.8.12). All $\Gamma_s$ on this circle produce the given $G_A$. For the given $G_A$, maximum output power is obtained with $\Gamma_L = \Gamma_{OUT}^*$, where $\Gamma_{OUT}$ is given by (3.2.6). This value of $\Gamma_L$ produces the transducer power gain $G_T = G_A$.

Since the constant available power gain circles and the constant noise figure circles are functions of $\Gamma_s$, they can be plotted together on the Smith chart and the trade-offs that result between gain and noise figure can be analyzed.

**Potentially Unstable Bilateral Case**

With a potentially unstable transistor the design procedure for a given $G_p$ is as follows:

1. For a given $G_p$, draw the constant operating power gain circle using (3.8.5) and (3.8.6), and also draw the output stability circle as discussed in Section 3.3 [i.e., see (3.3.7) and (3.3.8)]. Select a value of $\Gamma_L$ that is in the stable region and not too close to the stability circle.

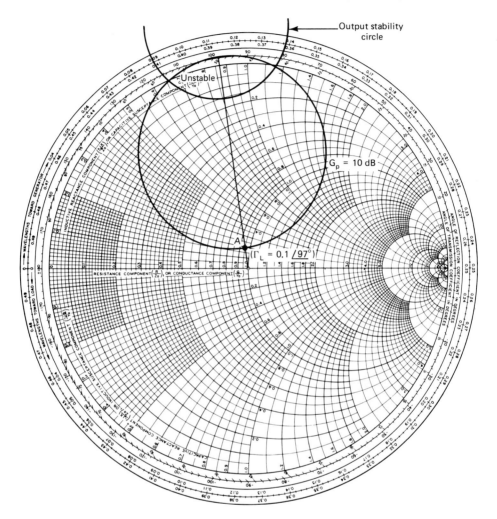

**Figure 3.8.2** Power and stability circles construction for Example 3.8.2.

2. Calculate $\Gamma_{IN}$ using (3.2.5) and determine if a conjugate match at the input is feasible. That is, draw the input stability circle as discussed in Section 3.3 [i.e., see (3.3.9) and (3.3.10)] and determine if $\Gamma_s = \Gamma_{IN}^*$ lies in the input stable region.

3. If $\Gamma_s = \Gamma_{IN}^*$ is not in the stable region, or in the stable region but very close to the input stability circle, the value of $\Gamma_s$ can be selected arbitrarily or a new value of $G_p$ selected. Of course, we must be careful when selecting $\Gamma_s$ arbitrarily since the value of $\Gamma_s$ affects the output power and the VSWR.

The values of $\Gamma_L$ and $\Gamma_s$ should not be too close to their respective stability circles, because oscillations might occur when the input and output circuits are tuned.

**Example 3.8.2**

The S parameters of a GaAs FET at $I_D = 50\% I_{DSS}$, $I_{DSS} = 10$ mA, $V_{DS} = 5$ V, and $f = 8$ GHz are

$$S_{11} = 0.5 \underline{|-180°}$$
$$S_{12} = 0.08 \underline{|30°}$$
$$S_{21} = 2.5 \underline{|70°}$$
$$S_{22} = 0.8 \underline{|-100°}$$

Design an amplifier with $G_p = 10$ dB.

**Solution.** First we must check the stability of the transistor. From (3.3.13) and (3.3.17) we obtain $K = 0.4$ and $\Delta = 0.223 \underline{|62.12°}$. Since $K < 1$, the GaAs FET is potentially unstable. The $G_{\text{MSG}}$ given by (3.6.11) is $G_{\text{MSG}} = 2.5/0.08 = 31.25$ or 14.9 dB.

In order to design for $G_p = 10$ dB (4.9 dB less than the $G_{\text{MSG}}$), the 10-dB operating power gain circle and the output stability circle must be calculated. The radius and center of the 10-dB power gain circle, from (3.8.5) and (3.8.6), are $R_p = 0.473$ and $C_p = 0.572 \underline{|97.2°}$. The radius and center of the output stability circle, from (3.3.7) and (3.3.8), are $r_L = 0.34$ and $C_L = 1.18 \underline{|97.2°}$.

The Smith chart in Fig. 3.8.2 shows the construction of the 10-dB operating power gain circle and the output stability circle. Since $|S_{11}| < 1$, the stable region is the region outside the output stability circle. $\Gamma_L$ is selected on the 10-dB power gain circle at location A, namely $\Gamma_L = 0.1 \underline{|97°}$ or $Z_L = 50(0.96 + j0.19) \, \Omega$.

For a conjugate match at the input, $\Gamma_s$ is given by $\Gamma_s = \Gamma_{\text{IN}}^* = 0.52 \underline{|179.32°}$ and we must determine if the value of $\Gamma_s$ is in the stable region. The radius and center of the input stability circle, from (3.3.9) and (3.3.10), are $r_s = 1.0$ and $C_s = 1.67 \underline{|171°}$, where the stable region is the region outside the input stability circle. Therefore, $\Gamma_s$ is a stable source reflection coefficient.

## 3.9 DC BIAS NETWORKS

It has been said that the least considered factor in microwave transistor amplifier design is the bias network [3.5]. While considerable effort is spent in designing for a given gain, noise figure, and bandwidth, little effort is spent in the dc bias network. The cost per decibel of microwave power gain or noise figure is high, and the designer cannot sacrifice the amplifier performance by having a poor dc bias design.

The purpose of a good dc bias design is to select the proper quiescent point and hold the quiescent point constant over variations in transistor parameters and temperature. A resistor bias network can be used with good results over moderate temperature changes. However, an active bias network is usually preferred for large temperature changes.

In the discussion that follows, we first consider the dc bias design for BJTs and then the bias design of GaAs FETs.

### BJT Bias Networks

At low frequencies, a bypassed emitter resistor is an important contributor to the quiescent-point stability. At microwave frequencies, the bypass capacitor, which is in parallel with the emitter resistor, can produce oscillations by making the input port unstable at some frequencies. Furthermore, an emitter resistor will degrade the noise performance of the amplifier. Therefore, in most microwave transistor amplifiers, especially in the gigahertz region, the emitter lead of the transistor is grounded.

At microwave frequencies the transistor parameters that are affected most by temperature are $I_{CBO}$, $h_{FE}$, and $V_{BE}$. The conventional reverse current $I_{CBO}$ (i.e., $I_{CBO}$ at low frequencies) doubles every 10°C rise in temperature. That is,

$$I_{CBO,T_2} = I_{CBO,T_1} 2^{(T_2 - T_1)/10}$$

where $I_{CBO,T_2}$ and $I_{CBO,T_1}$ are the values of $I_{CBO}$ at temperatures $T_2$ and $T_1$, respectively. The temperature $T_1$ is usually the temperature at which the manufacturer measures $I_{CBO}$. This temperature is usually 25°C.

A microwave transistor has a more complicated reverse current flow. The reverse current flow of a microwave transistor is composed of two components; one is the conventional $I_{CBO}$ and the other is a surface current, $I_s$, that flows across the top of the silicon lattice. The reverse current in a microwave transistor, which is referred to simply as $I_{CBO}$, increases at a rate much slower than the conventional $I_{CBO}$. A typical plot of the reverse current versus temperature for a microwave transistors is shown in Fig. 3.9.1. The conventional $I_{CBO}$ slope is also shown in the figure for comparison.

The base-to-emitter voltage $V_{BE}$ has a negative temperature coefficient, approximately given by

$$\frac{\Delta V_{BE}}{\Delta T} \approx -2 \times 10^3 \; \frac{\text{V}}{°\text{C}}$$

The dc value of the current gain $h_{FE}$ is defined as the value of the collector-to-base current at a constant value of $V_{CE}$. That is,

$$h_{FE} = \frac{I_C}{I_B}\bigg|_{V_{CE} = \text{CONSTANT}}$$

The dc value of $h_{FE}$ is typically found to increase linearly with temperature at the rate of 0.5%/°C.

In order to find the change in collector current as a function of temperature in a dc bias network, we first find the expression for the collector current valid for any temperature. Then, observing that the temperature sensitive parameters are $I_{CBO}$, $h_{FE}$, and $V_{BE}$, we can write

$$I_C = f(I_{CBO}, h_{FE}, V_{BE})$$

Sec. 3.9    DC Bias Networks

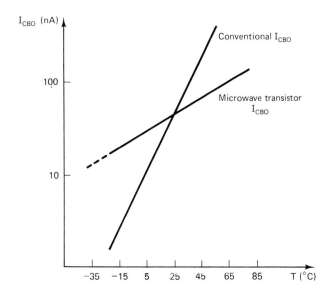

**Figure 3.9.1** Typical reverse current versus temperature for a microwave transistor.

and

$$\Delta I_C = \left(\frac{\Delta I_C}{\Delta I_{CBO}}\right)\bigg|_{\substack{\Delta h_{FE}=0 \\ \Delta V_{BE}=0}} \Delta I_{CBO} + \left(\frac{\Delta I_C}{\Delta h_{FE}}\right)\bigg|_{\substack{\Delta I_{CBO}=0 \\ \Delta V_{BE}=0}} \Delta h_{FE} + \left(\frac{\Delta I_C}{\Delta V_{BE}}\right)\bigg|_{\substack{\Delta I_{CBO}=0 \\ \Delta h_{FE}=0}} \Delta V_{BE} \tag{3.9.1}$$

Defining the stability factors as

$$S_i = \frac{\Delta I_C}{\Delta I_{CBO}}\bigg|_{\substack{\Delta h_{FE}=0 \\ \Delta V_{BE}=0}}$$

$$S_{h_{FE}} = \frac{\Delta I_C}{\Delta h_{FE}}\bigg|_{\substack{\Delta I_{CBO}=0 \\ \Delta V_{BE}=0}}$$

and

$$S_{V_{BE}} = \frac{\Delta I_C}{\Delta V_{BE}}\bigg|_{\substack{\Delta I_{CBO}=0 \\ \Delta h_{FE}=0}}$$

we can write (3.9.1) in the form

$$\Delta I_C = S_i \Delta I_{CBO} + S_{h_{FE}} \Delta h_{FE} + S_{V_{BE}} \Delta V_{BE} \tag{3.9.2}$$

For a given dc bias network, the stability factors can be calculated and (3.9.2) can be used to predict the variations of $I_C$ with temperature. In a design procedure, the maximum variation of $I_C$ in a temperature range can be selected and (3.9.2) can be used to find the required stability factors. In turn, the stability factors together with the $Q$-point location will fix the value of the resistors in the bias network.

Two grounded-emitter dc bias networks that can be used at microwave frequencies are shown in Fig. 3.9.2. The network in Fig. 3.9.2b produces lower values of resistance, and therefore is more compatible with thin- or thick-film resistor values.

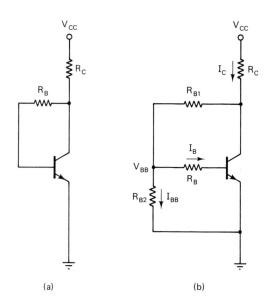

**Figure 3.9.2** (a) Voltage feedback bias network; (b) voltage feedback bias network with constant-base current source.

### Example 3.9.1

Design the dc bias network shown in Fig. 3.9.2b for $V_{CE} = 10$ V and $I_C = 10$ mA. Assume that $I_{CBO} = 0$, $V_{BE} = 0.7$ V, and $h_{FE} = 50$.

**Solution.** In this example we follow a procedure that results in good stability factors. Let the supply voltage $V_{CC}$ be 20 V. The base current ($I_B$) is

$$I_B = \frac{I_C}{h_{FE}} = \frac{10 \times 10^{-3}}{50} = 200 \; \mu\text{A}$$

Assuming $V_{BB}$ to be 2 V, we find that

$$R_B = \frac{V_{BB} - V_{BE}}{I_B} = \frac{2 - 0.7}{200 \times 10^{-6}} = 6.5 \; \text{k}\Omega$$

$R_{B2}$ is calculated assuming that $I_{BB} = 1$ mA (i.e., $I_{BB} = 5I_B$), namely

$$R_{B2} = \frac{V_{BB}}{I_{BB}} = \frac{2}{1 \times 10^{-3}} = 2 \; \text{k}\Omega$$

## Sec. 3.9  DC Bias Networks

$R_{B1}$ is obtained from

$$R_{B1} = \frac{V_{CE} - V_{BB}}{I_{BB} + I_B} = \frac{10 - 2}{(1 + 0.2) \times 10^{-3}} = 6.66 \text{ k}\Omega$$

and $R_C$ is obtained from

$$R_C = \frac{V_{CC} - V_{CE}}{I_C + I_{BB} + I_B} = \frac{20 - 10}{(10 + 1 + 0.2) \times 10^{-3}} = 893 \text{ }\Omega$$

The assumption $I_{BB} \gg I_B$ and $V_{BB} \approx 10\% V_{CC}$ produces good stability factors. It is left as an exercise to calculate the resulting stability factors.

At the lower microwave frequencies, the dc biasing network shown in Fig. 3.9.3, with a bypassed emitter resistor can be used. The bypassed emitter resistor provides excellent stability. For this network it is easy to show that

$$I_C = \frac{h_{FE}(V_{TH} - V_{BE})}{R_{TH} + (h_{FE} + 1)R_E} + \frac{(h_{FE} + 1)I_{CBO}(R_{TH} + R_E)}{R_{TH} + (h_{FE} + 1)R_E}$$

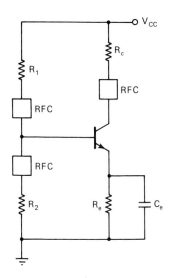

**Figure 3.9.3** A dc bias network with a bypassed emitter resistor.

where

$$V_{TH} = \frac{V_{CC} R_2}{R_1 + R_2}$$

and

$$R_{TH} = \frac{R_1 R_2}{R_1 + R_2}$$

The stability factors are

$$S_i = \frac{(h_{FE} + 1)(R_{TH} + R_E)}{R_{TH} + (h_{FE} + 1)R_E}$$

$$S_{h_{FE}} \approx \frac{I_{C1}}{h_{FE}} \frac{S_{i2}}{h_{FE,2}} \quad (3.9.3)$$

and

$$S_{V_{BE}} = \frac{-h_{FE}}{R_{TH} + (h_{FE} + 1)R_E}$$

In (3.9.3), $\Delta h_{FE} = h_{FE,2} - h_{FE}$ and $S_{i2}$ is the value of $S_i$ with $h_{FE} = h_{FE,2}$.

An active dc biasing network is shown in Fig. 3.9.4. A *pnp* BJT is used to stabilize the operating point of the microwave transistor. The bypass capacitors $C_1$ and $C_2$ are typically 0.01-$\mu$F disk capacitors. The radio frequency chokes (RFC) are typically made of two or three turns of No. 36 enameled wire on 0.1-in. air core. The operation of the network is as follows. If $I_{C2}$ tends to increase, the current $I_3$ increases and the emitter-to-base voltage of $Q_1$ ($V_{EB,1}$) decreases. The decrease of $V_{EB,1}$ decreases $I_{E1}$, which in turn decreases $I_{C2}$ and $I_{B2}$. The decrease in $I_{B2}$ and $I_{C2}$ produces the desired bias stability.

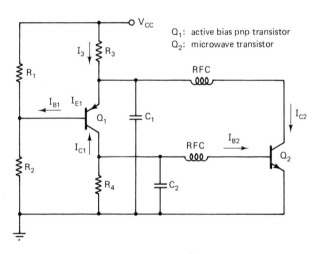

**Figure 3.9.4** Active bias network for a BJT.

The selection of the dc quiescent point for a BJT depends on the particular application. For low-noise and low-power applications, the quiescent point $A$ in Fig. 3.9.5 is recommended. At $A$, the BJT operates at low values of collector current. For low noise and higher power gain, the quiescent point at $B$ is recommended. For high output power, in class A operation, the quiescent point at $C$ is recommended. For higher output power and higher efficiency, the BJT is operated in class AB or B, using the quiescent point at $D$.

Sec. 3.9  DC Bias Networks    131

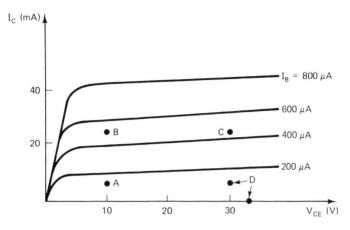

**Figure 3.9.5** Selection of the dc operating point.

### GaAs FET Bias Networks

The GaAs FETs can be biased in several ways. Five basic dc network configurations for GaAs FET amplifiers are shown in Fig. 3.9.6. [3.6]. The dc bias network in Fig. 3.9.6a requires a bipolar power source, while the networks in Figs. 3.9.6b to 3.9.6e require a unipolar supply. The column "How" in Fig. 3.9.6 indicates the polarity of the sources, as well as the sequence in which the voltages must be applied to prevent transient burnout of the GaAs FET device during turn-on. For example, in the dc bias network in Fig. 3.9.6a, if the drain is biased positive before the gate, the transistor will operate momentarily beyond its safe operating region. Therefore, the proper turn-on sequence is: first apply a negative bias to the gate (i.e., $V_G < 0$) and then apply the drain voltage ($V_D > 0$). One method to accomplish the previous turn-on procedure is to turn both sources at the same time and to include a long $RC$ time constant network in the $V_D$ supply and a short $RC$ time constant network in the negative supply $V_G$.

The bias networks in Figs. 3.9.6d and 3.9.6e use a source resistor. The source resistor provides automatic transient protection. However, the source resistor will degrade the noise-figure performance, and the source bypass capacitor can cause low-frequency oscillations.

The decoupling capacitors shown in Fig. 3.9.6 are sometimes shunted with zener diodes. The zener diodes provide additional protection against transients, reverse biasing, and overvoltage.

The dc bias network of a GaAs FET must provide a stable quiescent point. It is not difficult to show that the negative feedback resistor $R_s$ decreases the effect of variations of $I_D$ with respect to temperature and $I_{DSS}$.

The selection of the dc quiescent point in a GaAs FET depends on the particular application. Figure 3.9.7 shows typical GaAs FET characteristics with four quiescent points located at $A$, $B$, $C$, and $D$.

| Figure | | How | Amplifier characteristics | Power supply used |
|---|---|---|---|---|
| (a) $V_D = 5$ V $V_G = -2$ V 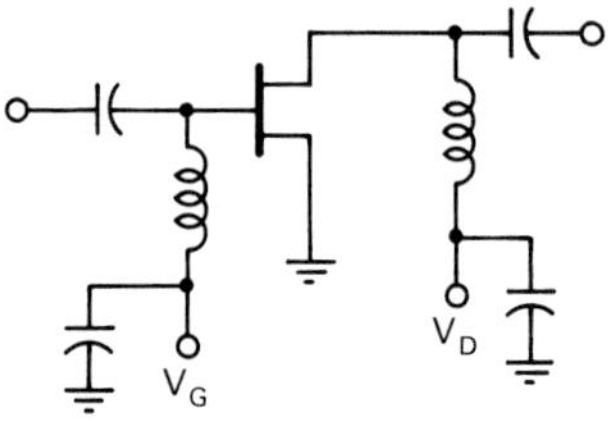 | | Apply $V_G$, then $V_D$ | Low noise High gain High power High efficiency | Bipolar: minimum source inductance |
| (b) $V_D = 7$ V $V_S = 2$ V 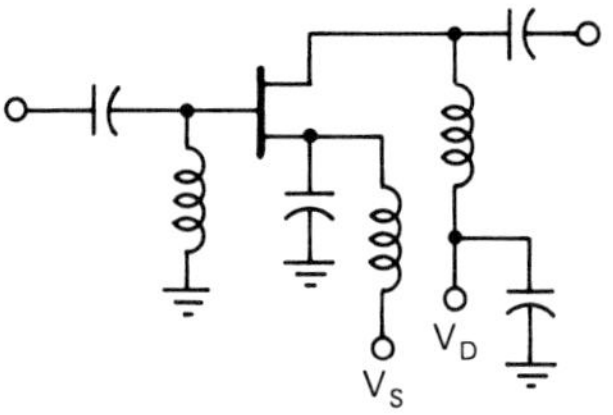 | | Apply $V_S$, then $V_D$ | [same as (a)] | Positive supply |
| (c) $V_G = -7$ V $V_S = -5$ V 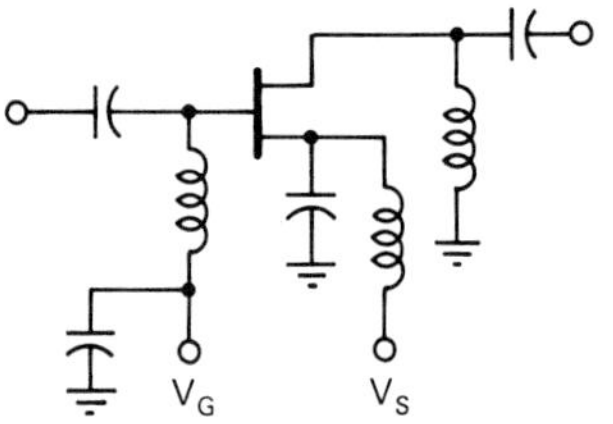 | | Apply $V_S$, then $V_G$ | [same as (a)] | Negative supply |
| (d) $V_D = 7$ V $V_S = 2$ V $= I_{DS} R_S$ 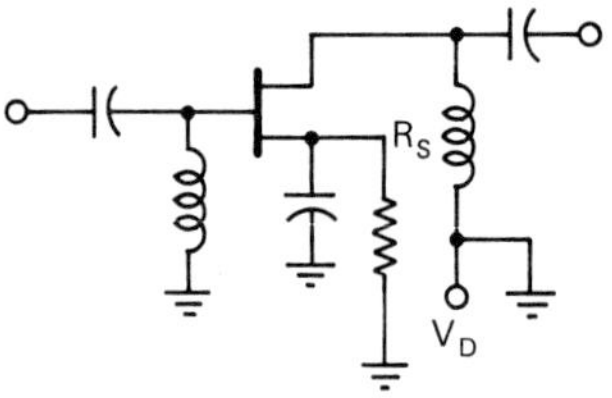 | | Apply $V_D$ | Low noise High gain High power Lower efficiency Gain easily adjusted by varying $R_S$ | Unipolar, incorporating $R_S$: automatic transient protection |
| (e) $V_G = -7$ V $V_S = -5$ V $= -I_{DS} R_S$ 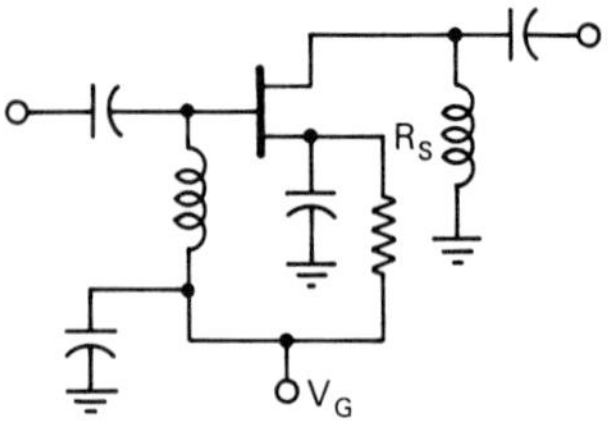 | | Apply $V_G$ | [same as (d)] | Negative unipolar, incorporating $R_S$ |

**Figure 3.9.6** Five basic dc bias networks. (From G. D. Vendelin [3.6]; reproduced with permission of Microwaves & RF.)

Sec. 3.9  DC Bias Networks  133

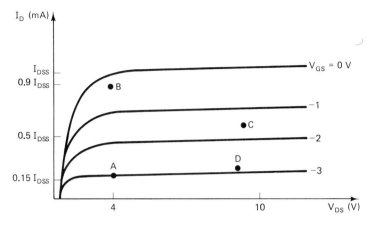

**Figure 3.9.7** Typical GaAs FET characteristics and recommended quiescent points.

For low-noise, low-power application, the quiescent point $A$ is recommended. At $A$, the FET operates at a low value of current (i.e., $I_{DS} \approx 0.15 I_{DSS}$).

For low noise and higher power gain, the recommended quiescent point is at $B$. The bias voltage remains the same as for point $A$, but the drain current is increased to $I_{DS} \approx 0.9 I_{DSS}$.

The GaAs FET output power level can be increased by selecting the quiescent point at $C$ with $I_{DS} \approx 0.5 I_{DSS}$. The quiescent point at $C$ maintains class A operation. For higher efficiency, or to operate the GaAs FET in class AB or B, the drain-to-source current must be decreased and the quiescent point $D$ is recommended.

An active bias network for a common-source GaAs FET is shown in Fig. 3.9.8.

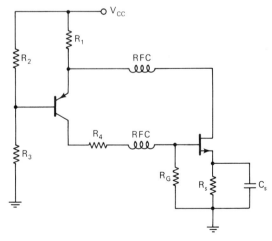

**Figure 3.9.8** Active bias for a common-source GaAs FET.

## PROBLEMS

**3.1.** Derive the expressions for $G_p$ and $G_A$ in (3.2.3) and (3.2.4) using Mason's rule.

**3.2.** (a) Show that $G_T \leq G_A$ and $G_T \leq G_p$. When is the equality sign satisfied?
   (b) Show that (3.2.3) can be obtained from (3.2.1) when $\Gamma_s = \Gamma_{IN}^*$, and (3.2.4) from (3.2.2) when $\Gamma_L = \Gamma_{OUT}^*$.

**3.3.** Prove that the maximum unilateral transducer power gain in (3.2.12) is obtained when $\Gamma_s = S_{11}^*$ and $\Gamma_L = S_{22}^*$.

**3.4.** Verify the stability circle equations in (3.3.5) and (3.3.6).

**3.5.** (a) Show that

$$|\Delta| \leq |S_{11}||S_{22}| + |S_{12}S_{21}|$$

and

$$|S_{11}||S_{22}| \leq |\Delta| + |S_{12}S_{21}|$$

Substitute these inequalities in (3.3.13) and verify that

$$(1 - |\Delta|)^2 > (|S_{11}|^2 - |S_{22}|^2)^2$$

Therefore, show that

$$B_1 B_2 > 0$$

and

$$B_1 + B_2 = 2(1 - |\Delta|^2)$$

where $B_1$ is given by (3.3.16) and

$$B_2 = 1 + |S_{22}|^2 - |S_{11}|^2 - |\Delta|^2$$

   (b) Show that the conditions $K > 1$ and $B_1 > 0$ are similar to $K > 1$ and $B_2 > 0$.
   (c) Show that the conditions $K > 1$ and $B_1 > 0$ are similar to the conditions $K > 1$ and $|\Delta| < 1$.
   (d) Verify that (3.3.14) and (3.3.15) are similar to $|\Delta| < 1$.

**3.6.** Prove that the necessary and sufficient conditions for unconditional stability are given by (3.3.13) to (3.3.15).

**3.7.** (a) Starting with (3.3.11) and (3.3.12), verify the conditions for unconditional stability.
   (b) The conditions for unconditional stability were analyzed by considering the values in the $\Gamma_s$ and $\Gamma_L$ plane that result in $|\Gamma_{IN}| < 1$ and $|\Gamma_{OUT}| < 1$. An alternative approach is to consider the values in the $\Gamma_{IN}$ and $\Gamma_{OUT}$ plane, where $|\Gamma_{IN}| < 1$ and $|\Gamma_{OUT}| < 1$, that results in $|\Gamma_s| < 1$ and $|\Gamma_L| < 1$. Using this approach, show that the plot of the $|\Gamma_s| = 1$ and $|\Gamma_L| = 1$ circles in the $\Gamma_{IN}$ and $\Gamma_{OUT}$ planes have radius given by

$$r_s = \frac{|S_{12}S_{21}|}{1 - |S_{11}|^2}$$

and

$$r_L = \frac{|S_{12}S_{21}|}{1 - |S_{22}|^2}$$

and centers given by

$$C_s = S_{22} + \frac{S_{12}S_{21}S_{11}^*}{1 - |S_{11}|^2}$$

and

$$C_L = S_{11} + \frac{S_{12}S_{21}S_{22}^*}{1 - |S_{22}|^2}$$

**3.8.** In each of the stability circle drawings shown in Fig. P3.8, indicate clearly the possible locations for a stable source reflection coefficient.

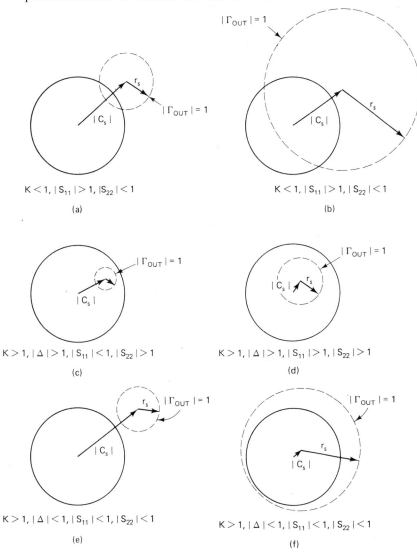

Figure P3.8

**3.9.** Verify the equation for the constant-gain circles in (3.4.6).

**3.10.** Show that the transducer power gain, when the input and output resistances to the transistor are $Z_o$, is given by $G_T = |S_{21}|^2$.

**3.11.** The scattering parameters for three different transistors are given below. Determine the stability in each case and in a potentially unstable case, draw the input and output stability circles.

(a) $S_{11} = 0.674 \underline{|-152°}$
$S_{12} = 0.075 \underline{|6.2°}$
$S_{21} = 1.74 \underline{|36.4°}$
$S_{22} = 0.6 \underline{|-92.6°}$

(b) $S_{11} = 0.385 \underline{|-55°}$
$S_{12} = 0.045 \underline{|90°}$
$S_{21} = 2.7 \underline{|78°}$
$S_{22} = 0.89 \underline{|-26.5°}$

(c) $S_{11} = 0.7 \underline{|-50°}$
$S_{12} = 0.27 \underline{|75°}$
$S_{21} = 5 \underline{|120°}$
$S_{22} = 0.6 \underline{|80°}$

**3.12.** Verify the equations for a simultaneous conjugate match in (3.6.5) and (3.6.6).

**3.13.** (a) Show that $|B_i/2C_i|^2 > 1$ is similar to $K^2 > 1$, and therefore that $|B_i/2C_i| > 1$ is similar to $|K| > 1$.

*Hint:* Prove and use the identities

$$|C_1|^2 = |S_{11} - \Delta S_{22}^*|^2 = |S_{12} S_{21}|^2 + (1 - |S_{22}|^2)(|S_{11}|^2 - |\Delta|^2)$$

and

$$|C_2|^2 = |S_{22} - \Delta S_{11}^*|^2 = |S_{12} S_{21}|^2 + (1 - |S_{11}|^2)(|S_{22}|^2 - |\Delta|^2)$$

(b) Analyze the solutions to (3.6.5) and (3.6.6) when $|K| > 1$ with $K$ negative.

**3.14.** Verify the equation for $G_{T,\max}$ in (3.6.10).

**3.15.** Show that the unconditionally stable criteria $|\Delta| < 1$ implies that $B_1$ and $B_2$ in (3.6.7) and (3.6.8) are greater than zero.

**3.16.** Show that for small $S_{12}$, $\Gamma_{Ms}$ and $\Gamma_{ML}$ are close to $S_{11}^*$ and $S_{22}^*$, respectively.

**3.17.** Verify the equation for the constant operating power gain circle in (3.8.4).

**3.18.** (a) Design a microwave transistor amplifier for $G_{TU,\max}$ using a BJT whose S parameters in a 50-Ω system at $V_{CE} = 10$ V, $I_C = 20$ mA, and $f = 1$ GHz are

$$S_{11} = 0.706 \underline{|-160°}$$
$$S_{12} = 0$$
$$S_{21} = 5.01 \underline{|85°}$$
$$S_{22} = 0.508 \underline{|-20°}$$

(b) Draw the constant-gain circles for $G_s = 2, 1, 0,$ and $-1$ dB.

**3.19.** The scattering parameters of a GaAs FET in a 50-Ω system are

$$S_{11} = 2.3 \underline{|-135°}$$
$$S_{12} = 0$$
$$S_{21} = 4 \underline{|60°}$$
$$S_{22} = 0.8 \underline{|-60°}$$

(a) Determine the unstable region in the Smith chart and construct the constant-gain circle for $G_s = 4$ dB.

(b) Design the input matching network for $G_s = 4$ dB with the greatest degree of stability.

(c) Draw the complete ac amplifier schematic.

**3.20.** Design a microwave transistor amplifier for $G_{T,\max}$ using a BJT whose $S$ parameters in a 50-Ω system at $V_{CE} = 10$ V, $I_C = 4$ mA, and $f = 750$ MHz are

$$S_{11} = 0.277 \underline{|-59°}$$
$$S_{12} = 0.078 \underline{|93°}$$
$$S_{21} = 1.92 \underline{|64°}$$
$$S_{22} = 0.848 \underline{|-31°}$$

(This problem is based on a design given in Ref. [3.7]).

**3.21.** Design a microwave transistor amplifier at $f = 750$ MHz to have $G_p = 10$ dB using the BJT in Problem 3.20. Also, determine the reflection coefficients for $G_{p,\max}$ and show that they are identical to $\Gamma_{Ms}$ and $\Gamma_{ML}$ in Problem 3.20.

**3.22.** At 2 GHz, a GaAs FET has the following $S$ parameters:

$$S_{11} = 0.7 \underline{|-65°}$$
$$S_{12} = 0.03 \underline{|60°}$$
$$S_{21} = 3.2 \underline{|110°}$$
$$S_{22} = 0.8 \underline{|-30°}$$

Determine the stability and design an amplifier with $G_p = 10$ dB.

**3.23.** (a) Show that the available power gain can be expressed in the form

$$G_A = \frac{|S_{21}|^2(1 - |\Gamma_s|^2)}{1 - |S_{22}|^2 + |\Gamma_s|^2(|S_{11}|^2 - |\Delta|^2) - 2\operatorname{Re}(C_1\Gamma_s)}$$

(b) Verify the constant available power gain circle equations in (3.8.9) to (3.8.12).
(c) Discuss the design procedure for a given $G_A$ in a potentially unstable bilateral case.

**3.24.** (a) Derive the stability factors for the dc bias networks in Fig. 3.9.2.
(b) Calculate the stability factors for the dc bias network in Example 3.9.1. If $h_{FE}$ changes from 50 to 150, what happens to the quiescent point?

**3.25.** Design the dc bias network shown in Fig. 3.9.3 for $V_{CE} = 6$ V, $I_C = 1$ mA, and $S_i = 5$. Assume that $h_{FE} = 100$, and $I_{CBO} = 1$ µA at 25°C. Calculate the resulting stability factors and find what happens to the operating point if the temperature increases to 75°C.

**3.26.** (a) In the network shown in Fig. 3.9.4, derive the expression for $I_C$ as a function of the network parameters.
(b) If $R_1 = 100$ kΩ, $R_2 = 200$ kΩ (potentiometer), $R_3 = 5$ kΩ (potentiometer), and $R_4 = 2.6$ kΩ, find the typical operating point of $Q_2$.

## REFERENCES

[3.1] K. Kurokawa, "Power Waves and the Scattering Matrix," *IEEE Transactions on Microwave Theory and Techniques*, March 1965.

[3.2] G. E. Bodway, "Two Port Power Flow Analysis Using Generalized Scattering Parameters," *Microwave Journal*, May 1967.

[3.3] D. Woods, "Reappraisal of the Unconditional Stability Criteria for Active 2-Port Networks in Terms of S Parameters," *IEEE Transactions on Circuits and Systems*, February 1976.

[3.4] T. T. Ha, *Solid State Microwave Amplifier Design*, Wiley-Interscience, New York, 1981.

[3.5] "Microwave Transistor Bias Considerations," Hewlett-Packard Application Note 944-1, April 1975.

[3.6] G. D. Vendelin, "Five Basic Bias Designs for GaAs FET Amplifiers," *Microwaves*, February 1978.

[3.7] W. H. Froehner, "Quick Amplifier Design with Scattering Parameters," *Electronics*, October 1967.

# 4

# NOISE, BROADBAND, AND HIGH-POWER DESIGN METHODS

## 4.1 INTRODUCTION

In Chapter 3, design methods for given stability and gain criteria were discussed. This chapter presents the basic principles involved in the design of low-noise, broadband, and high-power transistor amplifiers.

In some applications the design objective is for a minimum noise figure. Since a minimum noise figure and maximum power gain cannot be obtained simultaneously, constant noise figure circles, together with constant available power gain circles, can be drawn on the Smith chart, and reflection coefficients can be selected that compromise between the noise figure and gain performance. The trade-offs that result from noise considerations, stability, and gain are discussed in this chapter.

The noise performance of the GaAs FET is superior to that of the BJT above 4 GHz. A minimum noise figure in both BJTs and GaAs FETs is obtained at low collector or drain current.

The design philosophy in a broadband amplifier is to obtain flat gain over the prescribed range of frequencies. This can be obtained by the use of compensated matching networks, negative feedback, or balance amplifiers.

The small-signal $S$ parameters can be used in the design of microwave transistor amplifiers with linear power output (i.e., class A operation). However, the small-signal $S$ parameters are not useful in the design of large-output power amplifiers. In this case, large-signal impedance or reflection coefficient data as a function of output power and gain are needed.

## 4.2 NOISE IN TWO-PORT NETWORKS

In a microwave amplifier, even when there is no input signal, a small output voltage can be measured. We refer to this small output power as the *amplifier noise power*. The total noise output power is composed of the amplified noise input power plus the noise output power produced by the amplifier.

The model of a noisy two-port microwave amplifier is shown in Fig. 4.2.1. The noise input power can be modeled by a noisy resistor that produces thermal or Johnson noise. This noise is produced by the random fluctuations of the electrons due to thermal agitation. The rms value of the noise voltage $V_N$, produced by the noisy resistor $R_N$ over a frequency range $f_H - f_L$, is given by

$$V_N = \sqrt{4kTBR_N} \qquad (4.2.1)$$

where $k$ is Boltzmann's constant (i.e., $k = 1.374 \times 10^{-23}$ J/°K), $T$ is the resistor noise temperature, and $B$ is the noise bandwidth (i.e., $B = f_H - f_L$).

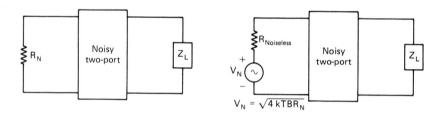

**Figure 4.2.1** Model of a noisy microwave amplifier.

Equation (4.2.1) shows that the thermal noise power depends on the bandwidth and not on a given center frequency. Such a distribution of noise is called *white noise*.

The maximum available noise power from $R_N$ is

$$P_N = \frac{V_N^2}{4R_N} = kTB \qquad (4.2.2)$$

**Example 4.2.1**

Calculate the noise voltage and maximum available noise power produced by a 2-MΩ resistor at a standard temperature ($T = 290°$K) in a 5-kHz bandwidth.

**Solution.** Using (4.2.1) and (4.2.2), the noise voltage and maximum available noise power are

$$V_N = \sqrt{4(1.374 \times 10^{-23})(290)(5 \times 10^3)(2 \times 10^6)} = 12.6 \ \mu\text{V}$$

and

$$P_N = \frac{(12.6 \times 10^{-6})^2}{4(2 \times 10^6)} = 19.9 \times 10^{-18} \text{ W}$$

## Sec. 4.2 Noise in Two-Port Networks

The noise figure ($F$) describes quantitatively the performance of a noisy microwave amplifier. The noise figure of a microwave amplifier is defined as the ratio of the total available noise power at the output of the amplifier to the available noise power at the output due to thermal noise from $R_N$. The noise figure can be expressed in the form

$$F = \frac{P_{N_o}}{P_{N_i} G_A} \quad (4.2.3)$$

where $P_{N_o}$ is the total available noise power at the output of the amplifier, $P_{N_i} = kTB$ is the available noise power due to $R_N$ in a bandwidth $B$, and $G_A$ is the available power gain.

Since $G_A$ can be expressed in the form

$$G_A = \frac{P_{S_o}}{P_{S_i}}$$

where $P_{S_o}$ is the available signal power at the output and $P_{S_i}$ is the available signal power at the input, then (4.2.3) can be written as

$$F = \frac{P_{S_i}/P_{N_i}}{P_{S_o}/P_{N_o}}$$

In other words, $F$ can also be defined as the ratio of the available signal-to-noise power ratio at the input to the available signal-to-noise power ratio at the output. A minimum noise figure is obtained by properly selecting the source reflection coefficient of the amplifier.

A model for the calculation of the noise figure of a two-stage amplifier is shown in Fig. 4.2.2. $P_{N_i}$ is the available input noise power, $G_{A1}$ and $G_{A2}$ are the available power gains of each stage, and $P_{n1}$ and $P_{n2}$ represent the noise power appearing at the output of amplifiers 1 and 2, respectively, due to the internal amplifier noise.

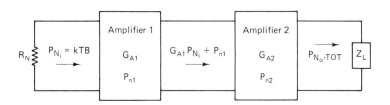

**Figure 4.2.2** Noise figure model of a two-stage amplifier.

The total available noise power at the output ($P_{N_o,\text{TOT}}$) is given by

$$P_{N_o,\text{TOT}} = G_{A2}(G_{A1} P_{N_i} + P_{n1}) + P_{n2}$$

Therefore, from (4.2.3) the noise figure of the two-stage amplifier is given by

$$F = \frac{P_{N_o,\text{TOT}}}{P_{N_i} G_{A1} G_{A2}} = 1 + \frac{P_{n1}}{P_{N_i} G_{A1}} + \frac{P_{n2}}{P_{N_i} G_{A1} G_{A2}}$$

or

$$F = F_1 + \frac{F_2 - 1}{G_{A1}} \qquad (4.2.4)$$

where

$$F_1 = 1 + \frac{P_{n1}}{P_{N_i} G_{A1}}$$

and

$$F_2 = 1 + \frac{P_{n2}}{P_{N_i} G_{A2}}$$

$F_1$ and $F_2$ are recognized as the individual noise figures of the first and second stages, respectively.

Equation (4.2.4) shows that the noise figure of the second stage is reduced by $G_{A1}$. Therefore, the noise contribution from the second stage is small if $G_{A1}$ is large, and can be significant if the gain $G_{A1}$ is low. It is not always important to minimize the first-stage noise if the gain reduction is too large. In fact, we can select a higher gain, even if $F_1$ is higher than the minimum noise figure of the first stage, such that a low value of $F$ is obtained. In a design, a trade-off between gain and noise figure is usually made.

## 4.3 CONSTANT NOISE FIGURE CIRCLES

The noise figure of a two-port amplifier is given by [4.1]

$$F = F_{\min} + \frac{r_n}{g_s} | Y_s - Y_o |^2 \qquad (4.3.1)$$

where $r_n$ is the equivalent normalized noise resistance of the two-port (i.e., $r_n = R_N/Z_o$), $Y_s = g_s + jb_s$ represents the source admittance, and $Y_o = g_o + jb_o$ represents that source admittance which results in the minimum noise figure, called $F_{\min}$.

We can express $Y_s$ and $Y_o$ in terms of the reflection coefficients $\Gamma_s$ and $\Gamma_o$, namely

$$Y_s = \frac{1 - \Gamma_s}{1 + \Gamma_s} \qquad (4.3.2)$$

## Sec. 4.3 Constant Noise Figure Circles

and

$$Y_o = \frac{1 - \Gamma_o}{1 + \Gamma_o} \qquad (4.3.3)$$

Substituting (4.3.2) and (4.3.3) into (4.3.1) results in the relation

$$F = F_{min} + \frac{4r_n|\Gamma_s - \Gamma_o|^2}{(1 - |\Gamma_s|^2)|1 + \Gamma_o|^2} \qquad (4.3.4)$$

Equation (4.3.4) depends on $F_{min}$, $r_n$, and $\Gamma_o$. These quantities are known as the *noise parameters* and are given by the manufacturer of the transistor or can be determined experimentally. The source reflection coefficient can be varied until a minimum noise figure is read in a noise figure meter. The value of $F_{min}$, which occurs when $\Gamma_s = \Gamma_o$, can be read from the meter, and the source reflection coefficient that produces $F_{min}$ can be determined accurately using a network analyzer. The noise resistance $r_n$ can be measured by reading the noise figure when $\Gamma_s = 0$, called $F_{\Gamma_s=0}$. Then, using (4.3.4) we obtain

$$r_n = (F_{\Gamma_s=0} - F_{min}) \frac{|1 + \Gamma_o|^2}{4|\Gamma_o|^2}$$

$F_{min}$ is a function of the device operating current and frequency, and there is one value of $\Gamma_o$ associated with each $F_{min}$. A typical plot of $F_{min}$ versus current for a BJT is illustrated in Fig. 4.3.1.

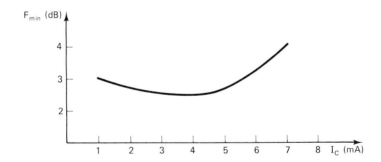

**Figure 4.3.1** Typical $F_{min}$ versus collector current measured at $V_{CE} = 10$ V and $f = 4$ GHz.

Equation (4.3.4) can be used to design $\Gamma_s$ for a given noise figure. For a given noise figure $F_i$, we define a noise figure parameter, called $N_i$, as

$$N_i = \frac{|\Gamma_s - \Gamma_o|^2}{1 - |\Gamma_s|^2} = \frac{F_i - F_{min}}{4r_n}|1 + \Gamma_o|^2 \qquad (4.3.5)$$

Equation (4.3.5) can be written as

$$(\Gamma_s - \Gamma_o)(\Gamma_s^* - \Gamma_o^*) = N_i - N_i|\Gamma_s|^2$$

or

$$|\Gamma_s|^2(1 + N_i) + |\Gamma_o|^2 - 2\,\text{Re}\,(\Gamma_s \Gamma_o^*) = N_i$$

If we now multiply both sides by $1 + N_i$, we obtain

$$|\Gamma_s|^2(1 + N_i)^2 + |\Gamma_o|^2 - 2(1 + N_i)\,\text{Re}\,(\Gamma_s \Gamma_o^*) = N_i^2 + N_i(1 - |\Gamma_o|^2)$$

or

$$\left|\Gamma_s - \frac{\Gamma_o}{1 + N_i}\right|^2 = \frac{N_i^2 + N_i(1 - |\Gamma_o|^2)}{(1 + N_i)^2} \qquad (4.3.6)$$

Equation (4.3.6) is recognized as a family of circles with $N_i$ as a parameter. The circles are centered at

$$C_{F_i} = \frac{\Gamma_o}{1 + N_i} \qquad (4.3.7)$$

with radii

$$R_{F_i} = \frac{1}{1 + N_i}\sqrt{N_i^2 + N_i(1 - |\Gamma_o|^2)} \qquad (4.3.8)$$

Equations (4.3.5), (4.3.7), and (4.3.8) show that when $F_i = F_{\min}$, then $N_i = 0$, $C_{F_{\min}} = \Gamma_o$, and $R_{F_{\min}} = 0$. That is, the center of the $F_{\min}$ circle is located at $\Gamma_o$ with zero radius. From (4.3.7), the centers of the other noise figure circles are located along the $\Gamma_o$ vector.

A typical set of constant noise figure circles is shown in Fig. 4.3.2. This set of curves show that $F_{\min} = 3$ dB is obtained when $\Gamma_s = \Gamma_o = 0.58\,\underline{/138°}$ and at point $A$, $\Gamma_s = 0.38\,\underline{/119°}$ produces $F_i = 4$ dB.

In a design there is always a difference between the designed noise figure and the measured noise figure of the final amplifier. This occurs because of the loss associated with the matching elements and the transistor noise figure variations from unit to unit. Typically, the noise figure difference can be from a fraction of a decibel to 1 dB in a narrowband design.

In the unilateral case, a set of $G_s$ constant-gain circles can be drawn in the Smith chart containing the noise figure circles. A typical plot for a GaAs FET is illustrated in Fig. 4.3.3. This plot shows the trade-offs that can be made between gain and noise figure in a design. Maximum gain and minimum noise figure cannot, in general, be obtained simultaneously. In Fig. 4.3.3, the maximum $G_s$ gain of 3 dB, obtained with $\Gamma_s = 0.7\,\underline{/110°}$, results in a noise figure of $F_i \approx 4$ dB; and the minimum noise figure $F_{\min} = 0.8$ dB, obtained with $\Gamma_s = 0.6\,\underline{/40°}$, results in a gain $G_s \approx -1$ dB.

### Sec. 4.3  Constant Noise Figure Circles

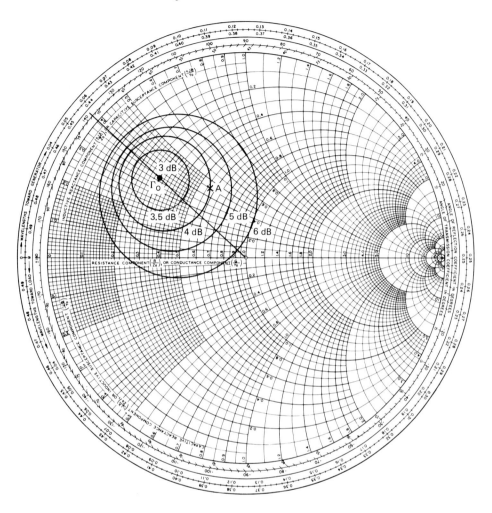

**Figure 4.3.2**  Typical constant noise figure circles in the $\Gamma_s$ plane.

### Example 4.3.1

The scattering and noise parameters of a BJT measured at a bias point for low-noise operation ($V_{CE} = 10$ V, $I_C = 4$ mA) at $f = 4$ GHz are

$$S_{11} = 0.552 \underline{|169°}$$
$$S_{12} = 0.049 \underline{|23°}$$
$$S_{21} = 1.681 \underline{|26°}$$
$$S_{22} = 0.839 \underline{|-67°}$$

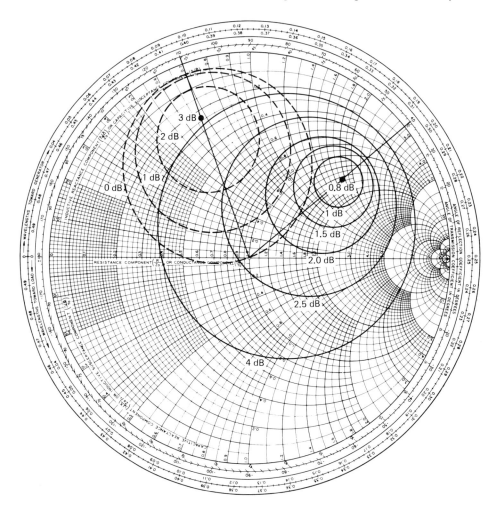

**Figure 4.3.3** Noise figure circles (solid curves) and $G_s$ constant-gain circles (dashed curves). The transistor is a GaAs FET with $V_{DS} = 4$ V, $I_{DS} = 12$ mA, and $f = 6$ GHz.

and

$$F_{min} = 2.5 \text{ dB}$$
$$\Gamma_o = 0.475 \underline{|166°}$$
$$R_N = 3.5 \text{ }\Omega$$

Design a microwave transistor amplifier to have a minimum noise figure. (This example is based on a design from Hewlett-Packard Application Note 967 [4.2].)

## Sec. 4.3 Constant Noise Figure Circles

**Solution.** The transistor is unconditionally stable at 4 GHz. A minimum noise figure of 2.5 dB is obtained with $\Gamma_s = \Gamma_o = 0.475 \underline{|166°}$. The constant noise figure circles in Fig. 4.3.4 for $F_i = 2.5$ to 3 dB were calculated using (4.3.5), (4.3.7), and (4.3.8). For example, the $F_i = 2.8$-dB circle was obtained as follows:

$$N_i = \frac{1.905 - 1.778}{4(3.5/50)} |1 + 0.475\underline{|166°}|^2 = 0.1378$$

$$C_{F_i} = \frac{0.475\underline{|166°}}{1 + 0.1378} = 0.417\underline{|166°}$$

and

$$R_{F_i} = \frac{1}{1 + 0.1378} \sqrt{(0.1378)^2 + 0.1378[1 - (0.475)^2]} = 0.312$$

Figure 4.3.4 shows that for this transistor $F_{\min}$ is not very sensitive to small variations in $\Gamma_s$ around $\Gamma_o$. In fact, the 2.6-dB constant-noise circle (i.e., a 0.1-dB increase in noise figure) results when $\Gamma_s$ changes in magnitude by 0.2 from its value at $\Gamma_o$.

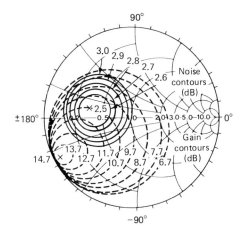

**Figure 4.3.4** Constant noise figure circles and available power gain circles. (From Ref. [4.2]; courtesy of Hewlett-Packard.)

The load reflection coefficient is selected to provide maximum gain for the lowest noise figure (i.e., with $\Gamma_s = \Gamma_o$) and, of course, for optimum VSWR at the output. Therefore,

$$\Gamma_L = \left(S_{22} + \frac{S_{12}S_{21}\Gamma_o}{1 - S_{11}\Gamma_o}\right)^* = 0.844\underline{|70.4°}$$

and the resulting gains are $G_T = G_A = 11$ dB and $G_p = 12.7$ dB.

The amplifier was designed, built, and tested by Hewlett-Packard [4.2]. The ac amplifier schematic is shown in Fig. 4.3.5. The input matching network was designed with a short-circuited stub and a quarter-wave transformer with $Z_o = 31.1 \ \Omega$. The output matching network was designed with a 0.61-cm microstrip line to provide soldering area, followed by a $\lambda/8$ short-circuited stub to tune out most of the susceptance component of $Y_1 = 1/Z_1$. Then, another series microstrip line followed by an

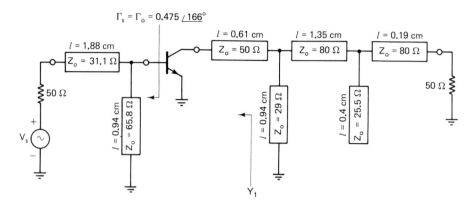

**Figure 4.3.5** Amplifier schematic. The microstrip lengths are given for $\varepsilon_{ff} = 1$ at $f = 4$ GHz. (From Ref. [4.2]; courtesy of Hewlett-Packard.)

open stub that provides some tuning capabilities was used. A final series microstrip line was used to obtain the match to 50 Ω.

Of course, the output matching network could have been designed differently. The form selected (see Fig. 4.3.5) provides flexibility for tuning by adjusting the lengths of the series lines (i.e., the $l = 1.35$-cm and $l = 0.19$-cm lines), by changing the width (i.e., the characteristic impedance) of the open-circuited stub and by modifying the lengths of the short-circuited stub.

The complete amplifier schematic and the microstrip board layout are shown in Fig. 4.3.6. The board material is Duroid ($\varepsilon_r = 2.23$, $h = 0.031$ in.).

The measured characteristics of the amplifier are shown in Fig. 4.3.7. Figure 4.3.7 shows that the amplifier performance is very good in the frequency range 3.7 to 4.2 GHz. The 3-dB bandwidth (see Fig. 4.3.7e) is 850 MHz, which corresponds to a 21% bandwidth.

It is easy to show that for this BJT, $K = 1.012$, $\Delta = 0.419 \lfloor 111.04°$, $G_{T,\max} = G_{A,\max} = 14.7$ dB, $\Gamma_{Ms} = 0.941 \lfloor -154°$, and $\Gamma_{ML} = 0.979 \lfloor 70°$. Since the available power gain with $\Gamma_s = \Gamma_o$ is $G_A = 11$ dB, a sacrifice in gain was needed to obtain optimum noise performance.

**Example 4.3.2**

The scattering and noise parameters of a GaAs FET measured at three different optimum bias settings at $f = 6$ GHz are:

Minimum Noise Figure ($V_{DS} = 3.5$ V, $I_{DS} = 15\% I_{DSS}$):

$$S_{11} = 0.674 \lfloor -152° \qquad F_{\min} = 2.2 \text{ dB}$$
$$S_{12} = 0.075 \lfloor 6.2° \qquad \Gamma_o = 0.575 \lfloor 138°$$
$$S_{21} = 1.74 \lfloor 36.4° \qquad R_N = 6.64 \text{ Ω}$$
$$S_{22} = 0.6 \lfloor -92.6°$$

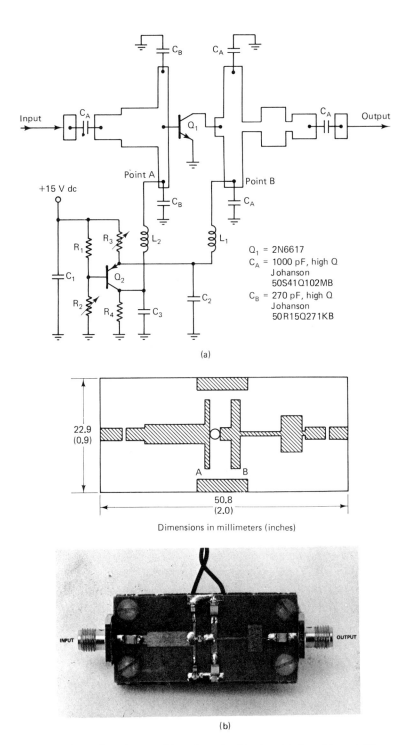

**Figure 4.3.6** (a) Complete amplifier schematic; (b) microstrip board layout. (From Ref. [4.2]; courtesy of Hewlett-Packard.)

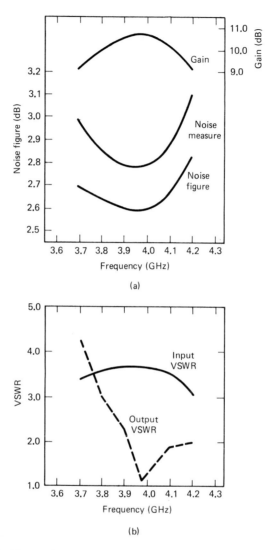

**Figure 4.3.7** Measured amplifier characteristics: (a) noise and gain performance; (b) input–output VSWR performance; (c) power output performance; (d) temperature performance at 4 GHz; (e) wideband gain performance. (From Ref. [4.2]; courtesy of Hewlett-Packard.)

*Linear Power Output* ($V_{DS} = 4$ V, $I_{DS} = 50\% I_{DSS}$):

$S_{11} = 0.641 \underline{\,-171.3°\,}$ $\qquad F_{min} = 2.9$ dB

$S_{12} = 0.057 \underline{\,16.3°\,}$ $\qquad \Gamma_o = 0.542 \underline{\,141°\,}$

$S_{21} = 2.058 \underline{\,28.5°\,}$ $\qquad R_N = 9.42\ \Omega$

$S_{22} = 0.572 \underline{\,-95.7°\,}$

## Sec. 4.3 Constant Noise Figure Circles

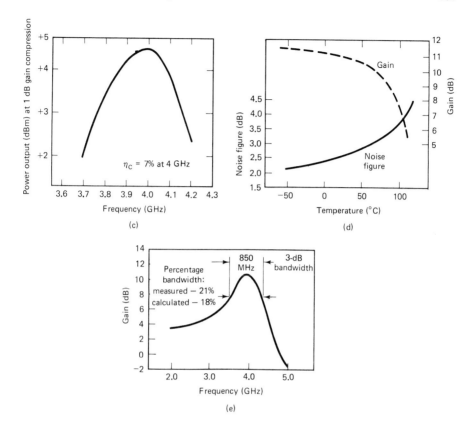

**Figure 4.3.7** (*continued*)

$Maximum\ Gain\ (V_{DS} = 4\ V, I_{DS} = 100\%I_{DSS})$:

$$S_{11} = 0.614 \underline{|-167.4°}$$
$$S_{12} = 0.046 \underline{|65°}$$
$$S_{21} = 2.187 \underline{|32.4°}$$
$$S_{22} = 0.716 \underline{|-83°}$$

Design a microwave transistor amplifier to have good ac performance. (This example is based on a design from Hewlett-Packard Application Note 970 [4.3].)

**Solution.** There are four ac performances that must be considered: noise figure, power gain, power output, and input and output VSWR. The linear power-output bias point ($V_{DS} = 4$ V, $I_{DS} = 50\%I_{DSS}$) provides a good compromise between the minimum noise figure and maximum gain. At this bias point Fig. 4.3.8 gives the noise, gain, and power parameters. The output power performance, measured at the 1-dB compression point, was experimentally measured and it is given in the figure. (See Section 4.7 for the definition of the 1-dB compression point.) The data for the output power were taken with an input power drive of 8.3 dBm.

| Noise Parameters | Gain Parameters | Power Parameters |
|---|---|---|
| $\Gamma_o = 0.542 \underline{|141°}$ | $\Gamma_{Ms} = 0.762 \underline{|177.3°}$ | $\Gamma_{PS} = 0.729 \underline{|166°}$ |
| $\Gamma_L = 0.575 \underline{|104.5°}$ | $\Gamma_{ML} = 0.718 \underline{|103.9°}$ | $\Gamma_{PL} = 0.489 \underline{|101°}$ |
| $F_{min} = 2.9$ dB | $F = 4.44$ dB | $F = 3.69$ dB |
| $G_A = 9.33$ dB | $G_{A,\,max} = 11.38$ dB | $G_p = 8.2$ dB |
| $P_{1dB} = 9.3$ dBm | $P_{1dB} = 13.4$ dBm | $P_{1dB} = 15.5$ dBm |

**Figure 4.3.8** Noise, gain, and power parameters at the linear power output bias ($V_{DS} = 4$ V, $I_D = 50\% I_{DSS}$).

The input VSWR with $\Gamma_s = \Gamma_{Ms}$ is 1, and the VSWR = 3.82 with $\Gamma_s = \Gamma_o$. In order to calculate the VSWR, we obtained $|\Gamma_a|$ (see Fig. 4.3.10a) and used (1.3.11), namely

$$\text{VSWR} = \frac{1 + |\Gamma_a|}{1 - |\Gamma_a|}$$

Other relations for calculating $|\Gamma_a|$ are given in Problem 4.6.

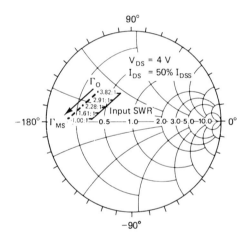

| $\Gamma_s$ | Mag./Ang. | $\Gamma_L$ Mag./Ang. | F (dB) | $G_A$ (dB) | Input VSWR | Output VSWR |
|---|---|---|---|---|---|---|
| $\Gamma_o$ — | 0.542 /141° | 0.575 /104° | 2.90 | 9.33 | 3.82:1 | 1.00:1 |
| | 0.572 /152° | 0.601 /105° | 2.97 | 10.04 | 2.91:1 | 1.00:1 |
| | 0.614 /160° | 0.627 /106° | 3.14 | 10.55 | 2.28:1 | 1.00:1 |
| | 0.678 /169° | 0.667 /105° | 3.57 | 11.10 | 1.61:1 | 1.00:1 |
| $\Gamma_{Ms}$ — | 0.762 /177° | 0.718 /104° | 4.44 | 11.38 | 1.00:1 | 1.00:1 |

**Figure 4.3.9** Trade-offs between noise figure, power gain, and VSWR. (From Ref. [4.3]; courtesy of Hewlett-Packard.)

## Sec. 4.3 Constant Noise Figure Circles

Figure 4.3.9 shows the noise figure, $G_A$ and input and output VSWR as the reflection coefficient is varied from $\Gamma_o$ to $\Gamma_{Ms}$, along a straight line, in the Smith chart. Figure 4.3.9 shows that a good compromise between noise figure, $G_A$, and VSWR is to use $\Gamma_s = 0.614 \underline{|160°}$ and $\Gamma_L = 0.627 \underline{|106°}$. The noise figure is increased by 0.24 dB from the minimum noise, but $G_A$ is increased by 1.22 dB and the input VSWR is improved by 40% (i.e., VSWR = 2.28). The ac schematic of the amplifier for the selected values of $\Gamma_s$ and $\Gamma_L$ is shown in Fig. 4.3.10a and the microstrip board layout is shown in Fig. 4.3.10b. The board material is Duroid ($\varepsilon_r = 2.23$, $h = 0.031$ in.). The measured characteristics of the amplifier are shown in Fig. 4.3.11.

In the last two examples the transistors were unconditionally stable. In a potentially unstable situation, we must check that the optimum noise reflection coefficient $\Gamma_s = \Gamma_o$ is in the stable region of the source stability circle. Once $\Gamma_s$ is selected, $\Gamma_L$ is selected for maximum gain (i.e., $\Gamma_L = \Gamma^*_{OUT}$), and again we must check that the value of $\Gamma_L$ is in the stable region of the load stability circle.

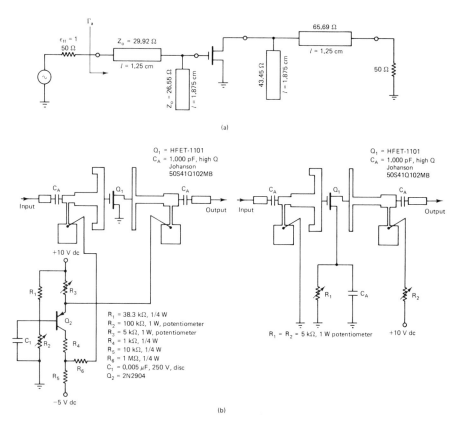

**Figure 4.3.10** (a) The ac schematic of the amplifier with $\varepsilon_{ff} = 1$; (b) microstrip layout with two different dc bias networks. (From Ref. [4.3]; courtesy of Hewlett-Packard.)

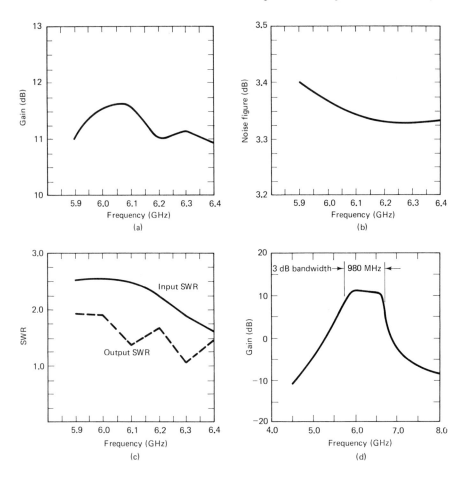

**Figure 4.3.11** Measured characteristics of the amplifier: (a) gain performance; (b) noise performance; (c) input–output VSWR performance; (d) wideband gain performance. (From Ref. [4.3]; courtesy of Hewlett-Packard.

## 4.4 BROADBAND AMPLIFIER DESIGN

The design of broadband amplifiers introduces new difficulties which require careful considerations. Basically, the design of a constant-gain amplifier over a broad frequency range is a matter of properly designing the matching networks, or the feedback network, in order to compensate for the variations of $|S_{21}|$ with frequency. The design specifications might require the use of rather sophisticated synthesis procedures in the design of the matching networks.

Some of the difficulties encountered in the design of a broadband amplifier are:

Sec. 4.4   Broadband Amplifier Design   155

1. The variations of $|S_{21}|$ and $|S_{12}|$ with frequency. Typically, $|S_{21}|$ decreases with frequency at the rate of 6 dB/octave and $|S_{12}|$ increases with frequency at the same rate. Typical variations of $|S_{21}|$, $|S_{12}|$, and $|S_{12}S_{21}|$ with frequency are illustrated in Fig. 1.9.7. The variations of $|S_{12}S_{21}|$ with frequency is important since the stability of the circuit depends on this quantity. It is in the flat region that we have to check the amplifier stability.
2. The scattering parameters $S_{11}$ and $S_{22}$ are also frequency dependent and their variations are significant over a broad range of frequencies.
3. There is a degradation of the noise figure and VSWR in some frequency range of the broadband amplifier.

Two techniques that are commonly used to design broadband amplifiers are (1) the use of compensated matching networks and (2) the use of negative feedback.

The technique of compensated matching networks involves mismatching the input and output matching networks to compensate for the changes with frequency of $|S_{21}|$. The matching networks are designed to give the best input and output VSWR. However, because of the broad bandwidth the VSWR will be optimum around certain frequencies, and a balanced amplifier design may be required.

The design of compensated matching networks can be done in analytical form with the help of the Smith chart. However, the use of a computer is usually required because of the complex analytical procedures. Of course, the use of a proper analytical procedure produces a starting design which can be optimized using computer-aided design (CAD) methods.

The matching networks can also be designed using network synthesis techniques. Passive network synthesis for the design of networks using lumped elements is well developed, and the techniques to implement the filter with microwave components are also well known [4.4, 4.5]. The microwave filters, typically, operate between two different impedances and must provide a prescribed insertion loss and bandwidth.

Insertion-loss synthesis techniques can be used to design impedance matching networks with prescribed responses. The synthesis process is a powerful tool when used in a CAD program. A good commercially available program is AMPSYN [4.6]. AMPSYN is a user-oriented interactive program for obtaining impedance matching networks with a desired frequency characteristic. AMPSYN synthesizes lumped elements matching networks and provides for transformations of the lumped design to approximate transmission-line equivalents.

An interesting method for broadband amplifier design suitable to CAD has been developed by Mellor [4.7]. The broadband design involves the use of an interstage matching network. The amplifier schematic is shown in Fig. 4.4.1.

Transistors $Q_1$ and $Q_2$ have a gain that decreases with increasing fre-

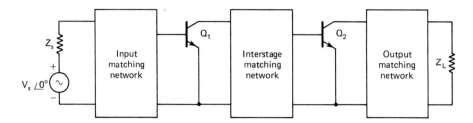

**Figure 4.4.1** Broadband amplifier schematic.

quency. The specifications for a good input and output match will require that the input and output matching networks have a constant gain over the frequency range of the amplifier (i.e., a flat frequency response). The interstage matching network must provide a gain having a positive slope with increasing frequency to compensate for the transistor roll-off and, therefore, to give an overall flat frequency response. The synthesis approach involves modeling the transistors with lumped elements and using an insertion-loss method to obtain the matching networks. The design of a broadband amplifier for a specific gain and noise figure requires, in general, the use of CAD techniques.

**Example 4.4.1**

The $S$ parameters of a BJT are given in Fig. 4.4.2. Design a broadband amplifier with a transducer power gain of 10 dB in the frequency range 300 to 700 MHz. (This example is based on a design from Hewlett-Packard Application Note 95-1 [4.8].)

| $f$ (MHz) | $S_{11}$ | $S_{21}$ | $S_{22}$ |
|---|---|---|---|
| 300 | $0.3 \angle -45°$ | $4.47 \angle 40°$ | $0.86 \angle -5°$ |
| 450 | $0.27 \angle -70°$ | $3.16 \angle 35°$ | $0.855 \angle -14°$ |
| 700 | $0.2 \angle -95°$ | $2.0 \angle 30°$ | $0.85 \angle -22°$ |

**Figure 4.4.2** Scattering parameters of a BJT.

**Solution.** The values in Fig. 4.4.2 show that

$$|S_{21}|^2 = 13 \text{ dB at } 300 \text{ MHz}$$
$$= 10 \text{ dB at } 450 \text{ MHz}$$
$$= 6 \text{ dB at } 700 \text{ MHz}$$

Therefore, in order to compensate for the variations of $|S_{21}|$, the matching networks must decrease the gain by 3 dB at 300 MHz, 0 dB at 450 MHz, and increase the gain by 4 dB at 700 MHz.

Sec. 4.4   Broadband Amplifier Design 157

For this transistor

$$G_{s,\max} = \frac{1}{1-|S_{11}|^2} = \begin{cases} 0.409 \text{ dB} & \text{at} \quad 300 \text{ MHz} \\ 0.329 \text{ dB} & \text{at} \quad 450 \text{ MHz} \\ 0.177 \text{ dB} & \text{at} \quad 700 \text{ MHz} \end{cases}$$

and little is to be gained by matching the source. Therefore, only the output matching network needs to be designed. Observe that $|S_{22}| \approx 0.85$ over the frequency range. Therefore,

$$G_{L,\max} = \frac{1}{1-|S_{22}|^2} = 5.6 \text{ dB}$$

and the gain of 4 dB from $G_L$ at 700 MHz is possible.

The output matching network is designed by plotting the constant-gain circles for $G_L = -3$ dB at 300 MHz, $G_L = 0$ dB at 450 MHz, and $G_L = 4$ dB at 700 MHz (see Fig. 4.4.3). The matching networks must transform the 50-$\Omega$ load to some point on the $-3$-dB circle at 300 MHz, to some point on the 0-dB circle at 450 MHz, and to some point on the 4-dB gain circle at 700 MHz. Of course, there are many matching networks that can perform the required transformation. The matching network selected is an ell network consisting of a shunt and series inductor combination (see Fig. 4.4.4).

The shunt inductor susceptance decreases with frequency and transforms the 50-$\Omega$ load along the constant-conductance circle as shown in Fig. 4.4.3. The series inductor reactance increases with frequency and transforms the parallel combination of 50 $\Omega$ and shunt inductance along a constant-resistance circle as shown in Fig. 4.4.3. Optimizing the values of $L_1$ and $L_2$ is a trial-and-error procedure. The graphical construction is illustrated in Fig. 4.4.3 and the final ac schematic of the amplifier is shown in Fig. 4.4.4.

The value of $L_1$ is obtained at 300 MHz (i.e., the lowest frequency), from Fig. 4.4.3, as

$$\frac{50}{j\omega L_1} = -j1.2$$

or

$$L_1 = 22.1 \text{ nH}$$

and the value of $L_2$ is obtained at 700 MHz (i.e., the highest frequency), from Fig. 4.4.3, as

$$\frac{j\omega L_2}{50} = j(3.2 - 0.4)$$

or

$$L_2 = 31.8 \text{ nH}$$

At the input, the direct connection of the 50-$\Omega$ source resistor to the base of the transistor results in $G_s = 0$ dB and an input VSWR smaller than 1.86 [i.e.,

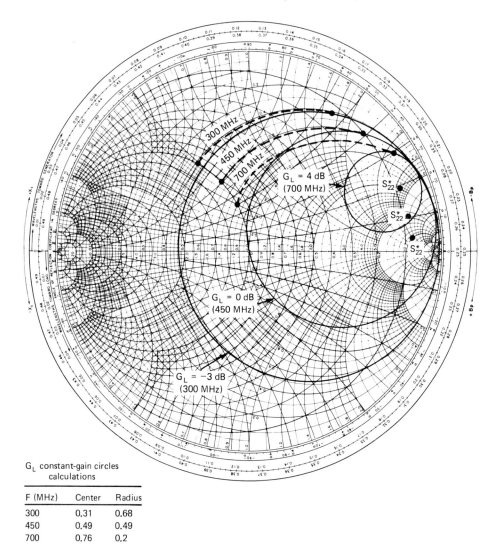

$G_L$ constant-gain circles calculations

| F (MHz) | Center | Radius |
|---|---|---|
| 300 | 0.31 | 0.68 |
| 450 | 0.49 | 0.49 |
| 700 | 0.76 | 0.2 |

**Figure 4.4.3**  Broadband design in the Smith chart.

$(1 + 0.3)/(1 - 0.3) = 1.86$]. The VSWR can be improved by matching the 50-Ω source to $S_{11}$ over the frequency band. The corresponding improvement in gain is small since $G_{s,\max} = 0.409$ dB at $f = 300$ MHz.

The design of compensated matching networks to obtain gain flatness results in impedance mismatching that can significantly degrade the input and output VSWR. The use of balanced amplifiers is a practical method for obtaining a broadband amplifier with flat gain and good input and output

Sec. 4.4    Broadband Amplifier Design 159

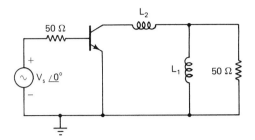

**Figure 4.4.4** The ac schematic of the broadband amplifier.

VSWR. The most popular arrangement of a balanced amplifier, shown in Fig. 4.4.5, uses two 3-dB hybrid couplers. A microstrip realization, shown in Fig. 4.4.6, uses an interdigitated structure which is known as the 3-dB *Lange coupler* [4.9].

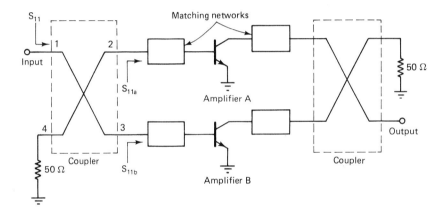

**Figure 4.4.5** Balanced amplifier configuration.

The input 3-dB coupler divides the input power equally between ports 2 and 3, and the output 3-dB coupler recombines the output signals from the amplifiers. The reflected signals at the input and output due to mismatching are coupled to the 50-Ω loads. It can be shown that the $S$ parameters of the $\lambda/4$, 3-dB Lange coupler are given by

$$|S_{11}| = 0.5|S_{11_a} - S_{11_b}|$$

$$|S_{21}| = 0.5|S_{21_a} + S_{21_b}|$$

$$|S_{12}| = 0.5|S_{12_a} + S_{12_b}|$$

$$|S_{22}| = 0.5|S_{22_a} - S_{22_b}|$$

If the two amplifiers are identical, then $S_{11} = 0$ and $S_{22} = 0$ and the gain $S_{21}$ (and also $S_{12}$) is equal to the gain of one side of the coupler. The bandwidth of

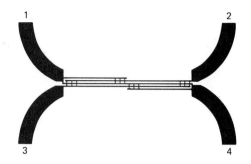

**Figure 4.4.6** A 3-dB Lange coupler. (From J. Lange [4.9]; copyright 1969, IEEE; reproduced with permission of IEEE.)

the balanced amplifier is limited by the bandwidth of the coupler (about 2 octaves).

The advantages of the balanced amplifier configuration are many:

1. The individual amplifiers can be designed for flat gain, noise figure, and so on (even if the individual amplifier VSWR is high), with the balanced amplifier input and output VSWR dependent on the coupler (i.e., ideally the VSWR is 1 if the amplifiers are identical).
2. A high degree of stability.
3. The output power is twice that obtained from the single amplifier.
4. If one of the amplifiers fails, the balanced amplifier unit will still operate with reduced gain.
5. Balanced amplifier units are easy to cascade with other units, since each unit is isolated by the coupler.

The disadvantages of the balanced amplifier configuration are that the unit uses two amplifiers, consumes more dc power, and is larger.

Negative feedback can be used in broadband amplifiers to provide a flat gain response and to reduce the input and output VSWR. It also controls the amplifier performance due to variations in the $S$ parameters from transistor to transistor. As the bandwidth requirements of the amplifier approach a decade of frequency, gain compensation based on matching networks only is very difficult, and negative feedback techniques are used. In fact, a microwave transistor amplifier using negative feedback can be designed to have very wide bandwidths (greater than 2 decades) with small gain variations (tenths of a decibel). On the minus side, negative feedback will degrade the noise figure and reduce the maximum power gain available from a transistor.

The most common methods of applying negative feedback are by the series and shunt resistor feedback configurations shown in Fig. 4.4.7. The coupling capacitors and the dc bias network have been omitted.

The following simple analysis illustrates the use of negative feedback. The BJT and GaAs FET can be represented by the equivalent circuit shown in Fig. 4.4.8 when the parasitic elements can be neglected (i.e., at low frequencies).

Sec. 4.4   Broadband Amplifier Design

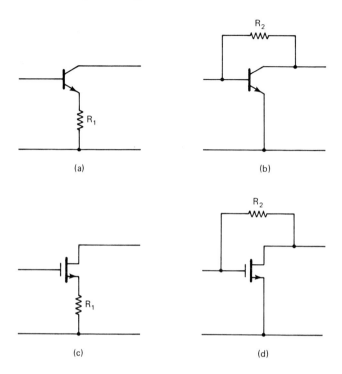

**Figure 4.4.7** (a) BJT with series feedback resistor; (b) BJT with shunt feedback resistor; (c) GaAs FET with series feedback resistor; (d) GaAs FET with shunt feedback resistor.

The resulting negative-feedback equivalent networks, including both series and shunt feedback, are shown in Fig. 4.4.9.

The admittance matrix for the network shown in Fig. 4.4.9b can be written in the form

$$\begin{bmatrix} i_1 \\ i_2 \end{bmatrix} = \begin{bmatrix} \dfrac{1}{R_2} & -\dfrac{1}{R_2} \\ \dfrac{g_m}{1+g_mR_1} - \dfrac{1}{R_2} & \dfrac{1}{R_2} \end{bmatrix} \begin{bmatrix} v_1 \\ v_2 \end{bmatrix}$$

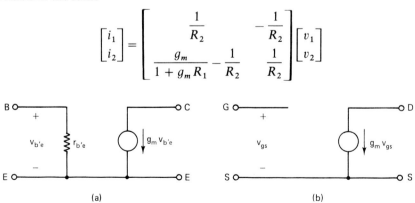

**Figure 4.4.8** (a) BJT equivalent network; (b) GaAs FET equivalent network.

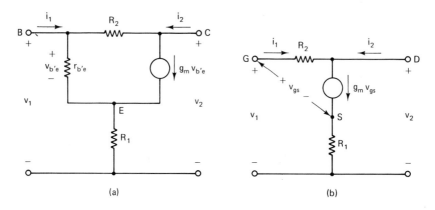

**Figure 4.4.9** (a) BJT negative-feedback model; (b) GaAs FET negative-feedback model.

and a similar matrix can be written for Fig. 4.4.9a. Using Fig. 1.8.1 to convert from $y$ parameters to $S$ parameters gives

$$S_{11} = S_{22} = \frac{1}{D}\left[1 - \frac{g_m Z_o^2}{R_2(1 + g_m R_1)}\right] \quad (4.4.1)$$

$$S_{21} = \frac{1}{D}\left(\frac{-2g_m Z_o}{1 + g_m R_1} + \frac{2Z_o}{R_2}\right) \quad (4.4.2)$$

and

$$S_{12} = \frac{2Z_o}{DR_2} \quad (4.4.3)$$

where

$$D = 1 + \frac{2Z_o}{R_2} + \frac{g_m Z_o^2}{R_2(1 + g_m R_1)}$$

From (4.4.1) the conditions $S_{11} = S_{22} = 0$ (i.e., input and output VSWR = 1) are satisfied when

$$1 + g_m R_1 = \frac{g_m Z_o^2}{R_2}$$

or

$$R_1 = \frac{Z_o^2}{R_2} - \frac{1}{g_m} \quad (4.4.4)$$

Substituting (4.4.4) into (4.4.2) and (4.4.3) gives

$$S_{21} = \frac{Z_o - R_2}{Z_o} \quad (4.4.5)$$

and

$$S_{12} = \frac{Z_o}{R_2 + Z_o}$$

Equation (4.4.4) shows that $S_{11} = S_{22} = 0$ can be satisfied, with positive values of $R_1$, if the transconductance is large.

A range of values that satisfies $S_{11} = S_{22} = 0$ can be found. The minimum transconductance [called $g_{m(min)}$] occurs when $R_1 = 0$ [i.e., $g_{m(min)} = R_2/Z_o^2$] and it follows from (4.4.5) that

$$g_{m(min)} = \frac{1 - S_{21}}{Z_o}$$

For example, if an amplifier has $|S_{21}|^2 = 10$ dB in a 50-Ω system, the minimum transconductance required is

$$g_{m(min)} = \frac{1 - (-3.16)}{50} = 83 \text{ mS}$$

and the required shunt feedback resistor is

$$R_2 = 83 \times 10^{-3}(50)^2 = 208 \text{ Ω}$$

Observe that (4.4.5) shows that

$$R_2 = Z_o(1 + |S_{21}|) \tag{4.4.6}$$

which is a well-known relation in feedback amplifiers.

When both $R_1$ and $R_2$ are used and $g_m$ has a high value, (4.4.4) shows that the minimum input and output VSWR is obtained when $R_1 R_2 \approx Z_o^2$. The high value of $g_m$ in a BJT makes them suitable for negative-feedback applications. However, most GaAs FETs have low values of $g_m$ and one must be careful when using them in negative-feedback configurations. Equation (4.4.5) shows that $S_{21}$ depends only on $R_2$ and not on the transistor parameters. Therefore, gain flattening can be achieved with negative feedback.

An important consideration in negative-feedback design is the phase of $S_{21}$. At low frequencies the phase of $S_{21}$ is close to 180°, and as the frequency increases (above $f_\beta$) the phase of $S_{21}$ varies rapidly. At some frequency the phase of $S_{21}$ is such that a portion of the output voltage is in phase with the input voltage (i.e., positive feedback). This problem can be solved by decreasing the feedback when the phase shift of $S_{21}$ approaches 90°. For example, in the case of shunt negative feedback (see Fig. 4.4.7), an inductor can be connected in series with $R_2$ such that after a certain frequency the negative feedback decreases in proportion to the $S_{21}$ roll-off.

The previous analysis, although based on a simplified model, can be used in a preliminary design. Then CAD methods can be used to calculate the $S$ parameters of the transistor with the feedback network connected, and to obtain the required $\Gamma_s$ and $\Gamma_L$ for optimum performance.

**Example 4.4.2**

Perform a preliminary analysis in the design of a BJT broadband amplifier having a transducer power gain of 10 dB from 10 to 1500 MHz. The $S$ parameters of the transistor (in a 50-$\Omega$ system) at 10 V, 4 mA, the associated $K$ factors, and $|S_{21}|^2$ in decibels (i.e., the transducer power gain in a 50-$\Omega$ system) are given in Fig. 4.4.10. (This example is based on a design from Ref. [4.10].)

| F (MHz) | $S_{11}$ Mag. | $S_{11}$ Ang. | $S_{21}$ Mag. | $S_{21}$ Ang. | $S_{12}$ Mag. | $S_{12}$ Ang. | $S_{22}$ Mag. | $S_{22}$ Ang. | $|S_{21}|^2$ (dB) | K |
|---|---|---|---|---|---|---|---|---|---|---|
| 10 | 0.95 | −2° | 7.35 | 174.6° | 0.003 | 84.3° | 1.01 | −1° | 17.3 | 0.11 |
| 100 | 0.92 | −11° | 7.15 | 168.0° | 0.007 | 79.0° | 0.99 | −4° | 17.1 | 0.18 |
| 250 | 0.87 | −28° | 6.83 | 154.5° | 0.015 | 69.2° | 0.96 | −10° | 16.7 | 0.29 |
| 500 | 0.78 | −54° | 6.28 | 135.0° | 0.026 | 54.0° | 0.90 | −18° | 16.0 | 0.42 |
| 750 | 0.69 | −78° | 5.67 | 123.0° | 0.033 | 41.4° | 0.84 | −25° | 15.1 | 0.53 |
| 1000 | 0.63 | −98° | 5.04 | 113.0° | 0.037 | 33.0° | 0.79 | −30° | 14.1 | 0.67 |
| 1250 | 0.60 | −114° | 4.42 | 99.9° | 0.038 | 29.3° | 0.77 | −33° | 13.0 | 0.81 |
| 1500 | 0.60 | −127° | 3.88 | 87.0° | 0.039 | 28.0° | 0.76 | −35° | 11.8 | 0.91 |

**Figure 4.4.10** $S$ parameters of the transistor, $K$ factors, and $|S_{21}|^2$ in decibels.

**Solution.** The transistor is certainly capable of providing a transducer power gain of 10 dB. However, since $K < 1$, the transistor is potentially unstable and a stability analysis must be performed. Also, observe that above 1250 MHz the phase of $S_{21}$ is less than 90°, and a portion of the output voltage is in phase with the input.

The input and output stability circles are given in Fig. 4.4.11. The analysis of the output stability circles show that a shunt resistor of 300 $\Omega$ at the output of the transistor provides stability. The resulting $S$ parameters for the network shown in Fig. 4.4.12a are given in Fig. 4.4.12b. The stability of the network in Fig. 4.4.12a is much improved.

| F (MHz) | $C_s$ Mag. | $C_s$ Ang. | $r_s$ | Stable Region | $C_L$ Mag. | $C_L$ Ang. | $r_L$ | Stable Region |
|---|---|---|---|---|---|---|---|---|
| 10 | 1.27 | −43° | 0.90 | Inside | 1.05 | 13° | 0.24 | Outside |
| 100 | 30.53 | 88° | 30.34 | Outside | 1.13 | 21° | 0.37 | Outside |
| 250 | 4.61 | 89° | 4.24 | Outside | 1.26 | 32° | 0.54 | Outside |
| 500 | 3.61 | 103° | 3.08 | Outside | 1.40 | 39° | 0.64 | Outside |
| 750 | 2.63 | 117° | 1.96 | Outside | 1.39 | 42° | 0.58 | Outside |
| 1000 | 2.26 | 129° | 1.46 | Outside | 1.41 | 44° | 0.53 | Outside |
| 1250 | 2.14 | 140° | 1.24 | Outside | 1.40 | 44° | 0.46 | Outside |
| 1500 | 1.99 | 150° | 1.04 | Outside | 1.39 | 45° | 0.42 | Outside |

**Figure 4.4.11** Stability circles locations.

Sec. 4.4  Broadband Amplifier Design                                                            165

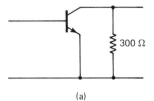

(a)

| F (MHz) | $S_{11}$ Mag. | $S_{11}$ Ang. | $S_{21}$ Mag. | $S_{21}$ Ang. | $S_{12}$ Mag. | $S_{12}$ Ang. | $S_{22}$ Mag. | $S_{22}$ Ang. | $|S_{21}|^2$ (dB) | K | $|\Delta|$ |
|---|---|---|---|---|---|---|---|---|---|---|---|
| 10   | 0.95 | −2°   | 6.30 | 174.7° | 0.003 | 84.4° | 0.72 | −1°  | 15.98 | 1.4  | 0.69 |
| 100  | 0.92 | −11°  | 6.13 | 168.3° | 0.006 | 79.3° | 0.71 | −4°  | 15.75 | 1.1  | 0.66 |
| 250  | 0.87 | −28°  | 5.88 | 155.2° | 0.013 | 69.9° | 0.69 | −10° | 15.38 | 0.94 | 0.61 |
| 500  | 0.79 | −53°  | 5.44 | 136.1° | 0.023 | 55.1° | 0.65 | −19° | 14.71 | 1.0  | 0.54 |
| 750  | 0.70 | −77°  | 4.94 | 124.5° | 0.029 | 42.9° | 0.61 | −26° | 13.88 | 1.2  | 0.45 |
| 1000 | 0.64 | −97°  | 4.42 | 114.7° | 0.032 | 34.7° | 0.57 | −32° | 12.90 | 1.4  | 0.37 |
| 1250 | 0.61 | −113° | 3.89 | 101.7° | 0.033 | 31.1° | 0.56 | −35° | 11.79 | 1.6  | 0.34 |
| 1500 | 0.60 | −126° | 3.42 | 88.8°  | 0.034 | 29.8° | 0.55 | −38° | 10.67 | 1.9  | 0.33 |

(b)

**Figure 4.4.12** (a) Stabilized transistor network; (b) the resulting S parameters.

The gain $|S_{21}|^2$ is reduced because the 300-Ω resistor dissipates some of the output power. Still, the network in Fig. 4.2.12a can easily provide the transducer power gain of 10 dB. The $S_{11}$ and $S_{22}$ parameters are large, showing that the input and output VSWR are poor. The phase of $S_{21}$ above 1250 MHz remains less than 90°.

The shunt negative-feedback resistor–inductor combination, shown in Fig. 4.4.13, can now be designed to provide a flat gain of 10 dB (i.e., $|S_{21}|^2 = 10$ dB or $|S_{21}| = 3.16$) with 50-Ω input and output impedances. The value of $R_2$ is calculated using (4.4.6), namely

$$R_2 = 50(1 + 3.16) = 208 \ \Omega$$

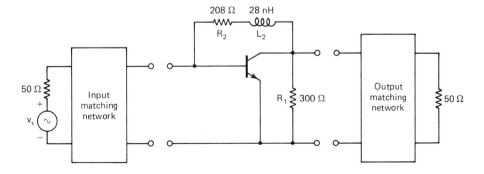

**Figure 4.4.13** The feedback network and the matching networks.

The value of $L_2$ is designed to provide negative feedback above 1200 MHz (i.e., to decrease the gain, so that the phase of $S_{21}$ in the feedback network remains above 90°). That is, the value of $L_2$ is selected from

$$R_2 = \omega L_2 |_{f=1200 \text{ MHz}} \quad \text{or} \quad L_2 = \frac{208}{2\pi(1200 \times 10^6)} = 28 \text{ nH}$$

The resulting $S$ parameters of the feedback network in Fig. 4.4.13 are given in Fig. 4.4.14. We can observe from Fig. 4.4.14a that negative feedback has reduced $S_{11}$ and $S_{22}$ considerably. The input VSWR is less than 2, except at 1500 MHz. Also, the gain $|S_{21}|^2$ in decibels in a 50-Ω system is close to the designed value of 10 dB over the frequency band.

In order to improve the input and output VSWR, and to flatten the gain, the feedback network elements (i.e., $R_2$ and $L_2$) can be varied using trial and error. Obviously, the number of calculations required are considerable and CAD methods are necessary.

This example is revisited in Appendix A, where CAD methods are used to optimize the feedback network design. In fact, it is shown that using $R_2 = 274.8$ Ω and $L_2 = 12.9$ nH results in the $S$ parameters given in Fig. 4.4.14b. From Fig. 4.4.14b it is observed that the gain flatness is improved.

| F (MHz) | $S_{11}$ Mag. | $S_{11}$ Ang. | $S_{21}$ Mag. | $S_{21}$ Ang. | $S_{12}$ Mag. | $S_{12}$ Ang. | $S_{22}$ Mag. | $S_{22}$ Ang. | $|S_{21}|^2$ (dB) | K | $|\Delta|$ |
|---|---|---|---|---|---|---|---|---|---|---|---|
| 10 | 0.08 | 17° | 2.68 | 176.7° | 0.184 | 1.0° | 0.04 | 140° | 8.58 | 1.25 | 0.50 |
| 100 | 0.07 | 17° | 2.67 | 175.5° | 0.182 | −2.9° | 0.07 | 113° | 8.53 | 1.26 | 0.48 |
| 250 | 0.08 | 11° | 2.73 | 171.5° | 0.180 | −9.6° | 0.14 | 93° | 8.74 | 1.23 | 0.48 |
| 500 | 0.10 | −5° | 3.03 | 163.7° | 0.172 | −20.4° | 0.26 | 77° | 9.62 | 1.14 | 0.51 |
| 750 | 0.12 | −46° | 3.37 | 156.6° | 0.154 | −32.6° | 0.35 | 62° | 10.56 | 1.11 | 0.53 |
| 1000 | 0.18 | −77° | 3.65 | 146.3° | 0.133 | −43.6° | 0.44 | 45° | 11.25 | 1.10 | 0.54 |
| 1250 | 0.28 | −93° | 3.79 | 128.9° | 0.112 | −54.4° | 0.56 | 26° | 11.58 | 1.08 | 0.56 |
| 1500 | 0.39 | −109° | 3.70 | 109.5° | 0.087 | −64.1° | 0.65 | 8° | 11.35 | 1.12 | 0.55 |

| F (MHz) | $S_{11}$ Mag. | $S_{11}$ Ang. | $S_{21}$ Mag. | $S_{21}$ Ang. | $S_{12}$ Mag. | $S_{12}$ Ang. | $S_{22}$ Mag. | $S_{22}$ Ang. | $|S_{21}|^2$ (dB) | K | $|\Delta|$ |
|---|---|---|---|---|---|---|---|---|---|---|---|
| 10.00 | 0.18 | 5° | 3.14 | 176.3° | 0.161 | 1.0° | 0.07 | 20° | 9.95 | 1.22 | 0.52 |
| 100.00 | 0.18 | −4° | 3.11 | 173.0° | 0.159 | −1.9° | 0.08 | 32° | 9.86 | 1.23 | 0.51 |
| 250.00 | 0.17 | −22° | 3.10 | 165.5° | 0.156 | −6.8° | 0.11 | 45° | 9.83 | 1.25 | 0.50 |
| 500.00 | 0.18 | −52° | 3.17 | 153.7° | 0.148 | −14.0° | 0.17 | 49° | 10.01 | 1.26 | 0.49 |
| 750.00 | 0.21 | −87° | 3.20 | 145.8° | 0.134 | −21.2° | 0.21 | 44° | 10.11 | 1.33 | 0.47 |
| 1000.00 | 0.25 | −110° | 3.19 | 137.4° | 0.120 | −26.6° | 0.25 | 36° | 10.06 | 1.40 | 0.44 |
| 1250.00 | 0.30 | −120° | 3.14 | 124.2° | 0.109 | −31.2° | 0.34 | 25° | 9.95 | 1.45 | 0.44 |
| 1500.00 | 0.36 | −128° | 3.06 | 109.5° | 0.098 | −35.7° | 0.43 | 15° | 9.72 | 1.49 | 0.45 |

**Figure 4.4.14** (a) $S$ parameters of the feedback network with $R_2 = 208$ Ω and $L_2 = 28$ nH; (b) $S$ parameters of the feedback network with $R_2 = 274.8$ Ω and $L_2 = 12.9$ nH.

## Sec. 4.4  Broadband Amplifier Design

Further performance improvements can be obtained with the addition of input and output matching networks, and using CAD methods to optimize the overall design for a flat transducer power gain of 10 dB with good input and output VSWR. The optimization and final design are discussed in Appendix A.

A matching network produces the desired match at one frequency and matching degradation occurs at the other frequencies. Fano [4.11] has derived a complete set of integrals that predict the gain–bandwidth restrictions for lossless matching networks terminated in an arbitrary load impedance. The derivations of the results are covered in Fano's paper and only the appropriate results will be given. For the network shown in Fig. 4.4.15a, the best $\Gamma$ that can be achieved over a frequency range is restricted by the integral

$$\int_0^\infty \ln \left| \frac{1}{\Gamma} \right| d\omega \leq \frac{\pi}{RC} \qquad (4.4.7)$$

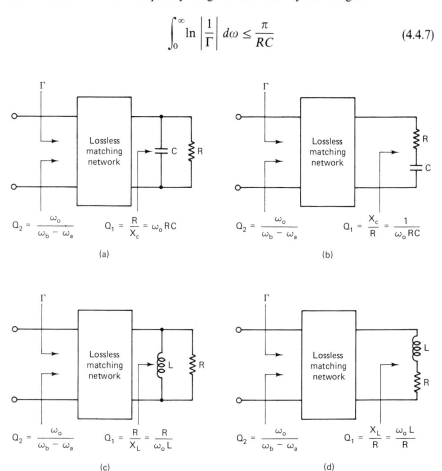

**Figure 4.4.15** Network topologies used in the calculations of the gain–bandwidth limitations.

Equation (4.4.7) expresses the fact that the area under the curve $\ln|1/\Gamma|$ cannot be greater than $\pi/RC$. Therefore, if matching is required over a certain bandwidth, it can be obtained at the expense of less power transfer.

The best utilization of the area under the curve $\ln|1/\Gamma|$ is obtained when $|\Gamma|$ is constant over the frequency range $\omega_a$ to $\omega_b$ and equal to 1 outside that range. This situation is illustrated in Fig. 4.4.16, and it follows from (4.4.7) that

$$|\Gamma| = \Gamma_x = e^{-\pi/(\omega_b - \omega_a)RC}$$

or

$$\Gamma_x = e^{-\pi(Q_2/Q_1)} \qquad (4.4.8)$$

where

$$Q_1 = \frac{R}{X_c}$$

and

$$Q_2 = \frac{\omega_o}{\omega_b - \omega_a}$$

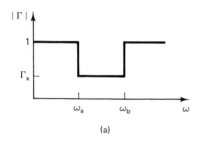
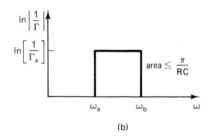

**Figure 4.4.16** Optimum values of $|\Gamma|$.

Equation (4.4.8) gives the best ideally achievable $\Gamma_x$ that can be obtained in the band $\omega_a$ to $\omega_b$ with no power transfer outside the band. Although a matching network satisfying the requirements above cannot be obtained in practice, the relation (4.4.8) can be used as a guideline for the best $\Gamma_x$.

The expression (4.4.8) can also be used for the networks shown in Figs. 4.4.15b to 4.4.15d when the appropriate definition of $Q_1$ and $Q_2$ are used. These are given in the figures.

**Example 4.4.3**

Over the frequency range $f_a = 500$ MHz to $f_b = 900$ MHz, find the best $\Gamma_x$ that can be achieved in the network shown in Fig. 4.4.17.

Sec. 4.5    Amplifier Tuning

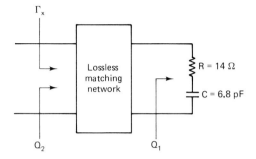

**Figure 4.4.17** Calculation of $\Gamma_x$.

**Solution.** At $f_o = 700$ MHz we obtain

$$Q_1 = \frac{1}{\omega_o RC} = \frac{1}{2\pi(700 \times 10^6)14(6.8 \times 10^{-12})} = 2.39$$

and

$$Q_2 = \tfrac{7}{4} = 1.75$$

Then, from (4.4.8), the value of $\Gamma_x$ is

$$\Gamma_x = e^{-\pi(1.75/2.39)} = 0.1$$

The normalized load impedance of the network in Fig. 4.4.17 (i.e., $z = 0.28 - j468 \times 10^6/f$) is plotted in Fig. 4.4.18 over the frequency range $f_a$ to $f_b$. Also, the region $\Gamma_x < 0.1$ for the ideally achievable match is shown shaded.

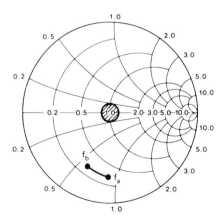

**Figure 4.4.18** Best achievable match $\Gamma_x$.

## 4.5 AMPLIFIER TUNING

The input reflection coefficient in the bilateral case is a function of the output reflection coefficient, and vice versa. Therefore, $\Gamma_{\text{IN}}$ varies with output tuning and $\Gamma_{\text{OUT}}$ varies with input tuning.

After an amplifier is built, tuning or alignment is necessary in order for the amplifier to provide optimum performance. The tuning is usually done by performing minor changes and adjusting the components of the matching networks.

An amplifier is easy to tune when the ratios of the fractional changes in $\Gamma_{IN}$ due to $\Gamma_L$, and $\Gamma_{OUT}$ due to $\Gamma_s$, are small.

The input reflection tuning factor $\delta_{IN}$ is defined as

$$\delta_{IN} = \left| \frac{d\Gamma_{IN}/\Gamma_{IN}}{d\Gamma_L/\Gamma_L} \right|$$

which in terms of the two-port network $S$ parameters can be written as

$$\delta_{IN} = \frac{|S_{21}||S_{12}||\Gamma_L|}{|1 - S_{22}\Gamma_L||S_{11} - \Delta\Gamma_L|} \qquad (4.5.1)$$

In practice, a value of $\delta_{IN} < 0.3$ produces good tunability. Equation (4.5.1) shows that $\delta_{IN}$ can be zero under some circumstances. That is, $\delta_{IN} = 0$ when $S_{12} = 0$, which occurs when the unilateral assumption can be made. In this case, the output tuning does not affect the input. Also, $\delta_{IN} = 0$ when $\Gamma_L = 0$, which is a very specific value of $\Gamma_L$ that probably degrades the gain and noise performance of the amplifier. Of course, $\delta_{IN} = 0$ when $S_{21} = 0$, that is, when there is no power gain. The derivation for the output tunability factor $\delta_{OUT}$ is left as an exercise.

Equation (4.5.1) can be solved for $\Gamma_L$ in terms of $\delta_{IN}$, namely

$$|\Gamma_L| = \left| \alpha \pm \left| \alpha^2 - \frac{S_{11}}{S_{22}\Delta} \right|^{1/2} \right| \qquad (4.5.2)$$

where

$$\alpha = \frac{\Delta + S_{11}S_{22} + S_{12}S_{21}\delta_{IN}^{-1}}{2S_{22}\Delta} \qquad (4.5.3)$$

The value of $|\Gamma_L|$ obtained from (4.5.2) and (4.5.3) for a given $\delta_{IN}$ is, in general, different from the value that produces maximum power gain or optimum noise performance. Therefore, this value of $\Gamma_L$ produces good tunability but mismatches the amplifier.

## 4.6 BANDWIDTH ANALYSIS

The conditions for a conjugate match at the input and output ports are satisfied at one frequency. One reason the output power varies with frequency is the frequency dependence of the matching networks. However, the most important factor that limits the frequency response is the variations of the transistor $S$ parameters with frequency.

## Sec. 4.6  Bandwidth Analysis

The input port, under conjugate matched conditions, is shown in Fig. 4.6.1a. A conjugate match means that $\Gamma_s = \Gamma_{IN}^*$ or $Y_s = Y_{IN}^*$. With

$$Y_{IN} = G_{s,M} + jB_{s,M}$$

and

$$Y_s = G_{s,M} - jB_{s,M}$$

the network in Fig. 4.6.1a can be represented by the *RLC* network shown in Fig. 4.6.1b. The values $G_{s,M}$ and $B_{s,M}$ represent the conductance and susceptance obtained under conjugate match conditions.

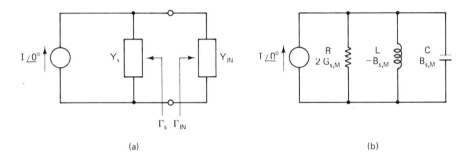

**Figure 4.6.1** Equivalent networks of the input port under conjugate match conditions.

The input inherent bandwidth, $(BW)_{IN}^i$, is the bandwidth obtained under conjugate matched conditions where the matched terminations are determined by the *S* parameters of the two-port device. The input inherent bandwidth is given by

$$(BW)_{IN}^i = \frac{f_o}{Q_{IN}} \quad \text{Hz} \quad (4.6.1)$$

where $f_o$ is the frequency at which the conjugate matched values were obtained and $Q_{IN}$ is the "que" of the equivalent input network. The value of $Q_{IN}$, for the network in Fig. 4.6.1b, can be expressed in different forms, namely

$$Q_{IN} = \omega_o RC = \frac{R}{\omega_o L} \quad (4.6.2)$$

where $\omega_o = 2\pi f_o = 1/\sqrt{LC}$. Substituting (4.6.2) into (4.6.1), we obtain

$$(BW)_{IN}^i = \frac{2f_o G_{s,M}}{|B_{s,M}|} \quad (4.6.3)$$

where $R = 1/2G_{s,M}$ and $|B_{s,M}| = \omega_o C = 1/\omega_o L$. Similarly, the output inherent bandwidth is given by

$$(\text{BW})^i_{\text{OUT}} = \frac{2f_o G_{s,M}}{|B_{L,M}|} \tag{4.6.4}$$

where $Y_L = Y^*_{\text{OUT}}$.

**Example 4.6.1**

In the microwave transistor amplifier of Example 3.7.1 we found that for a simultaneous conjugate match at $f = 6$ GHz, $\Gamma_{Ms} = 0.762 \underline{|177.3°|}$, and $\Gamma_{ML} = 0.718 \underline{|103.9°|}$. Calculate the amplifier bandwidth limitation due to the matching networks.

**Solution.** The admittances $Y_{\text{IN}} = Y^*_s$ and $Y_{\text{OUT}} = Y^*_L$ associated with $\Gamma_{Ms}$ and $\Gamma_{ML}$ are

$$Y_{\text{IN}} = (144 + j24.6) \times 10^{-3} \text{ S}$$

and

$$Y_{\text{OUT}} = (8.28 + j23.8) \times 10^{-3} \text{ S}$$

The equivalent network at the input port is illustrated in Fig. 4.6.1. The equivalent network for the output port is similar. From $Y_{\text{IN}}$ and $Y_{\text{OUT}}$, it follows that $G_{s,M} = 144 \times 10^{-3}$, $B_{s,M} = 24.6 \times 10^{-3}$, $G_{L,M} = 8.28 \times 10^{-3}$, and $B_{L,M} = 23.8 \times 10^{-3}$. Therefore, from (4.6.3) and (4.6.4),

$$(\text{BW})^i_{\text{IN}} = \frac{2(6 \times 10^9)144 \times 10^{-3}}{24.6 \times 10^{-3}} = 70.2 \text{ GHz}$$

and

$$(\text{BW})^i_{\text{OUT}} = \frac{2(6 \times 10^9)8.28 \times 10^{-3}}{23.8 \times 10^{-3}} = 4.17 \text{ GHz}$$

Since $(\text{BW})^i_{\text{IN}} \gg (\text{BW})^i_{\text{OUT}}$, the bandwidth limitations due to the matching networks are determined by $(\text{BW})^i_{\text{OUT}}$.

The broad bandwidth $(\text{BW})^i_{\text{OUT}}$ cannot be obtained in practice because of the transistor S-parameter variations with frequency. In fact, the overall bandwidth of the amplifier, as shown in Fig. 4.3.11d, is 980 MHz.

The inherent bandwidth of either the input or output port can be decreased by increasing the $Q$ of the network. From (4.6.2), we can increase $Q$ by increasing the capacitance or decreasing the inductance of the network. When $Y_{\text{IN}}$ has a capacitive susceptance, the bandwidth is decreased by adding capacitance, and when $Y_{\text{IN}}$ has an inductive susceptance, the bandwidth is decreased by adding inductance.

Consider the case where $Y_{\text{IN}}$ has a capacitive susceptance. The admittance $Y_{\text{IN}}$ is given by

$$Y_{\text{IN}} = G_{\text{IN},M} + jB_{\text{IN},M}$$

Sec. 4.6    Bandwidth Analysis

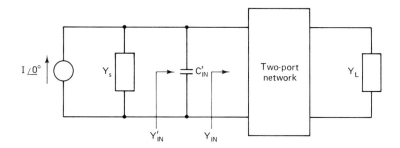

**Figure 4.6.2** Increasing $Q$ by adding the capacitance $C'_{IN}$.

If we add to the input port the capacitor $C'_{IN}$, shown in Fig. 4.6.2, the admittance $Y'_{IN}$ is given by

$$Y'_{IN} = Y_{IN} + j\omega_o C'_{IN}$$

and a conjugate match requires that $Y_s = (Y'_{IN})^*$. Therefore,

$$G_{s,M} = G_{IN,M}$$

and

$$B_{s,M} = -(B_{IN,M} + \omega_o C'_{IN})$$

The input port bandwidth is

$$(BW)_{IN} = \frac{2f_o G_{s,M}}{B_{IN,M} + \omega_o C'_{IN}}$$

which can be solved for $C'_{IN}$, to obtain

$$C'_{IN} = \frac{B_{IN,M}}{\omega_o}\left[\frac{(BW)^i_{IN}}{(BW)_{IN}} - 1\right] \quad (4.6.5)$$

where (4.6.3.) was used.

Equation (4.6.5) gives the value of the additional capacitance required to obtain the bandwidth $(BW)_{IN}$. Similarly, for the output network the capacitance required to produce a bandwidth $(BW)_{OUT}$ is

$$C'_{OUT} = \frac{B_{OUT,M}}{\omega_o}\left[\frac{(BW)^i_{OUT}}{(BW)_{OUT}} - 1\right]$$

When $Y_{IN}$ has an inductive susceptance, the bandwidth is decreased by adding the inductor $L'_{IN}$ shown in Fig. 4.6.3. It follows that the value of inductance required to obtain the bandwidth $(BW)_{IN}$ is

$$L'_{IN} = \frac{1}{\omega_o |B_{IN,M}|\left[\frac{(BW)^i_{IN}}{(BW)_{IN}} - 1\right]}$$

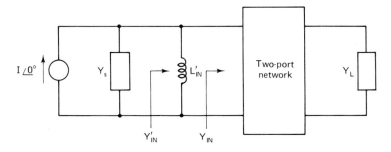

**Figure 4.6.3** Increasing $Q$ by adding the inductance $L'_{IN}$.

In the output port, the value of inductance required to obtain $(BW)_{OUT}$ is

$$L'_{OUT} = \frac{1}{\omega_o |B_{OUT,M}| \left[\frac{(BW)^i_{OUT}}{(BW)_{OUT}} - 1\right]}$$

The previous methods of adding capacitance or inductance to narrowband the amplifier response does not affect the original simultaneous conjugate matched calculations.

The overall bandwidth of $n$ identical single tuned networks is related to the bandwidth of one stage, $(BW)_1$, by the relation

$$(BW)_n = (BW)_1 \sqrt{2^{1/n} - 1} \qquad (4.6.6)$$

The factor

$$\sqrt{2^{1/n} - 1}$$

is called the *bandwidth reduction factor*. In the case of two single tuned networks (i.e., $n = 2$), (4.6.6) gives

$$(BW)_2 = (BW)_1 (0.644)$$

## 4.7 HIGH-POWER AMPLIFIER DESIGN

Thus far we have presented design techniques, based on the small-signal $S$ parameters of transistors, for maximum or arbitrary power gain, low noise, and broadband amplifiers. The small-signal $S$ parameters are not useful for power amplifier design because power amplifiers usually operate in nonlinear regions. The small-signal $S$ parameters can be used in large-signal amplifiers operating in class A (i.e., linear output power). However, for class AB, B, or C the small-signal $S$ parameters are not suitable for design purposes.

A set of large-signal $S$ parameters is needed to characterize the transistor for power applications. Unfortunately, the measurement of large-signal $S$ parameters is difficult and is not properly defined. Therefore, an alternative set of

### Sec. 4.7 High-Power Amplifier Design

large-signal parameters is needed to characterize the transistor. This can be done by providing information of source and load reflection coefficients as a function of output power and gain, especially the measurement of the source and load reflection coefficients, together with the output power, when the transistor is operated at its 1-dB gain compression point. The listing of the 1-dB compression point data is used to specify the power-handling capabilities of the transistor.

The 1-dB gain compression point (called $G_{1dB}$) is defined as the power gain where the nonlinearities of the transistor reduces the power gain by 1 dB over the small-signal linear power gain. That is,

$$G_{1dB}(dB) = G_o(dB) - 1 \qquad (4.7.1)$$

where $G_o(dB)$ is the small-signal linear power gain in decibels. Since the power gain is defined as

$$G_p = \frac{P_{OUT}}{P_{IN}}$$

or

$$P_{OUT}(dBm) = G_p(dB) + P_{IN}(dBm)$$

we can write the output power at the 1-dB gain compression point, called $P_{1dB}$, as

$$P_{1dB}(dBm) = G_{1dB}(dB) + P_{IN}(dBm) \qquad (4.7.2)$$

Substituting (4.7.1) into (4.7.2) gives

$$P_{1dB}(dBm) - P_{IN}(dBm) = G_o(dB) - 1 \qquad (4.7.3)$$

Equation (4.7.3) shows that the 1-dB gain compression point is that point at which the output power minus the input power in dBm is equal to the small-signal power gain minus 1 dB.

A typical plot of $P_{OUT}$ versus $P_{IN}$ which illustrates the 1-dB gain compression point is shown in Fig. 4.7.1. Observe the linear output power characteristics for power levels between the minimum detectable signal output power ($P_{o,mds}$) and $P_{1dB}$. The dynamic range (DR), shown in Fig. 4.7.1, is that range where the amplifier has a linear power gain. The dynamic range is limited at low power levels by the noise level. An input signal ($P_{i,mds}$) is detectable only if its output power level ($P_{o,mds}$) is above the noise power level.

The thermal noise power level of a two-port, with noise figure $F$, is given by

$$P_{N_o} = kTBG_A F$$

Observing that $kT = -174$ dBm (or $kTB = -114$ dBm/MHz at $T = 290°K$) and assuming that the minimum detectable input signal is $X$ decibels above

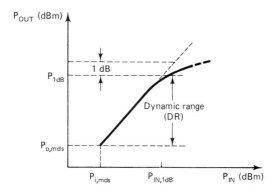

**Figure 4.7.1** The 1-dB gain compression point and the dynamic range of microwave amplifiers.

thermal noise, we can write

$$P_{i,\text{mds}} = -174 \text{ dBm} + 10 \log B + F(\text{dB}) + X(\text{dB}) \tag{4.7.4}$$

and

$$P_{o,\text{mds}} = -174 \text{ dBm} + 10 \log B + F(\text{dB}) + X(\text{dB}) + G_A(\text{dB}) \tag{4.7.5}$$

A typical value of $X(\text{dB})$ is 3 dB.

As previously discussed, a power transistor can be described in terms of the large-signal source and load reflection coefficients required to produce a given output power and gain. Of course, these parameters are functions of frequency and bias conditions. For example, in a GaAs FET the 1-dB gain compression point is usually measured at a drain-to-source voltage and gate-to-source voltage that optimizes the output power. From Fig. 3.9.7, this is usually at $I_{DS} = 50\% I_{DSS}$.

A typical set of power reflection coefficients is shown in Fig. 4.7.2a. The values of $\Gamma_{SP}$ and $\Gamma_{LP}$ denoted by points are the source and load power reflection coefficients for maximum output power. The values of $\Gamma_{SP}$ and $\Gamma_{LP}$ are given for $f = 4$ GHz to $f = 12$ GHz. Figure 4.7.2b illustrates typical output power contours as a function of the load reflection coefficient. For this transistor $P_{1\text{dB}} = 19$ dBm and $G_{1\text{dB}} = 6$ dB. The 18-dBm and 17-dBm output power contours are also shown. The input was conjugately matched at all times.

A typical measuring system for large-signal parameters is illustrated in Fig. 4.7.3. The transistor under test is placed in a measuring setup where the dc bias and ac input signal level can be varied. The output tuning stubs are adjusted until the power meter $C$ measures a given power level and the input tuning stubs are adjusted for zero reflected power (read at power meter $B$). The power meter $A$ reads the incident power, and the power gain (at given output power level) can be obtained. Since there is no reflected power, the input port is conjugately matched, and the output impedance is that impedance required to produce the output power read at power meter $C$.

## Sec. 4.7 High-Power Amplifier Design

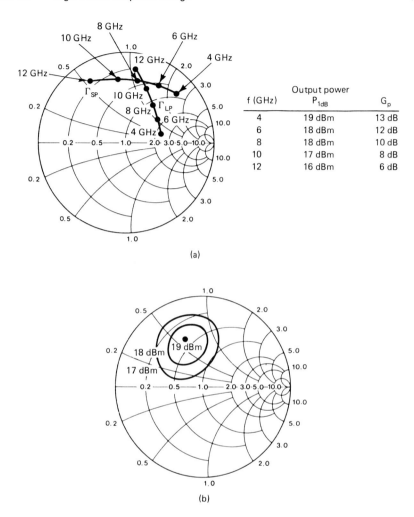

| f (GHz) | Output power $P_{1dB}$ | $G_p$ |
|---|---|---|
| 4  | 19 dBm | 13 dB |
| 6  | 18 dBm | 12 dB |
| 8  | 18 dBm | 10 dB |
| 10 | 17 dBm | 8 dB  |
| 12 | 16 dBm | 6 dB  |

**Figure 4.7.2** (a) Typical large-signal reflection coefficients. (b) Typical output power contours as a function of $\Gamma_{LP}$ for a GaAs FET at $f = 10$ GHz, $V_{DS} = 10$ V, and $I_D = 50\% I_{DSS}$. For this transistor the optimum output power is 19 dBm at 1-dB gain compression and $G_{1dB} = 6$ dB.

The transistor can then be disconnected from the test setup and the impedance at the reference planes A and B is measured with a network analyzer. These measurements produce the values of $\Gamma_{SP}$ and $\Gamma_{LP}$ for a given output power and gain. Of course, $\Gamma_{SP}$ and $\Gamma_{LP}$ are functions of output power level, frequency, and bias conditions.

A source of distortion in power amplifiers is that caused by intermodulation products. When two or more sinusoidal frequencies are applied to a nonlinear amplifier, the output contains additional frequency components

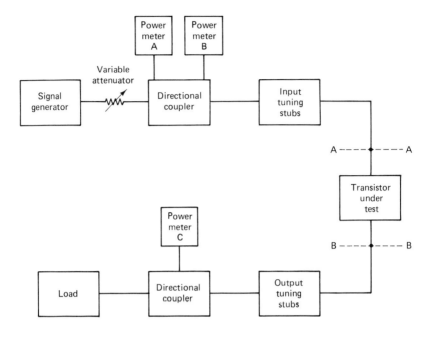

**Figure 4.7.3** Measuring system for large-signal parameters.

called *intermodulation products*. For example, if two sinusoidal signals

$$v(t) = A \cos 2\pi f_1 t + A \cos 2\pi f_2 t \qquad (4.7.6)$$

are applied to a nonlinear amplifier whose output voltage can be represented by the power series

$$v_o(t) = \alpha_1 v(t) + \alpha_2 v^2(t) + \alpha_3 v^3(t) \qquad (4.7.7)$$

the output signal will contain frequency components at dc, $f_1, f_2, 2f_1, 2f_2, 3f_1, 3f_2, f_1 \pm f_2, 2f_1 \pm f_2$, and $2f_2 \pm f_1$. The frequencies $2f_1$ and $2f_2$ are the second harmonics, $3f_1$ and $3f_2$ are the third harmonics, $f_1 \pm f_2$ are the second-order intermodulation products (since the sum of the $f_1$ and $f_2$ coefficients is 2), and $2f_1 \pm f_2$ and $2f_2 \pm f_1$ are the third-order intermodulation products (since the sum of the $f_1$ and $f_2$ coefficients is 3). The input and output power spectra, from (4.7.6) and (4.7.7), are shown in Fig. 4.7.4.

Figure 4.7.4 shows that the third-order intermodulation products at $2f_1 - f_2$ and $2f_2 - f_1$ are very close to the fundamental frequencies $f_1$ and $f_2$ and fall within the amplifier bandwidth, producing distortion in the output.

If we measure the third-order intermodulation product output power ($P_{2f_1 - f_2}$) versus the input power at $f_1$ ($P_{f_1}$), the graph shown in Fig. 4.7.5 results. The third-order intercept point (called $P_{IP}$) is defined as the point where $P_{f_1}$ and $P_{2f_1 - f_2}$ intercept, when the two-port is assumed to be linear. Observe that the slope of $P_{f_1}$ is 1 and that of $P_{2f_1 - f_2}$ is 3. This occurs because

Sec. 4.7  High-Power Amplifier Design

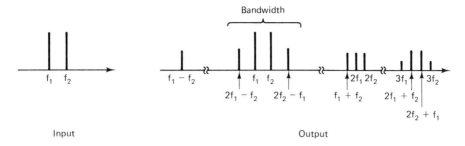

Figure 4.7.4 Input and output power spectrum.

for the assumed $v_o(t)$ in (4.7.7) the power of the third-order intermodulation product is proportional to the cube of the input signal amplitude $A$. The power $P_{IP}$ is a theoretical level. However, it is a useful quantity to estimate the third-order intermodulation products at different power levels.

For the three-term series in (4.7.7), it can be shown analytically and experimentally that the third-order intercept point is approximately 10 dB above the 1-dB gain compression point. That is,

$$P_{IP}(\text{dBm}) = P_{1\text{dB}}(\text{dBm}) + 10 \text{ dB} \tag{4.7.8}$$

Also, it can be shown that

$$2P_{2f_1-f_2} = 3P_{f_1} - 2P_{IP}$$

or

$$P_{f_1} - P_{2f_1-f_2} = \tfrac{2}{3}(P_{IP} - P_{2f_1-f_2}) \tag{4.7.9}$$

The spurious free dynamic range ($\text{DR}_f$) of an amplifier (see Fig. 4.7.5) is defined as the range $P_{f_1} - P_{2f_1-f_2}$, when $P_{2f_1-f_2}$ is equal to the minimum detectable output signal. Therefore, from (4.7.5) and (4.7.9),

$$\text{DR}_f = \tfrac{2}{3}(P_{IP} - P_{o,\text{mds}})$$
$$= \tfrac{2}{3}[P_{IP} + 174 \text{ dBm} - 10 \log B - F(\text{dB}) - X(\text{dB}) - G_A(\text{dB})]$$

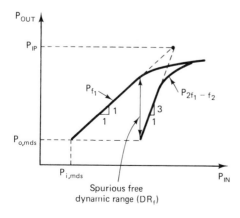

Figure 4.7.5 Third-order intercept point.

### Example 4.7.1

An amplifier has an available power gain of 40 dB, 500-MHz bandwidth, noise figure of 7 dB, and a 1-dB gain compression point of 25 dBm. Calculate DR and $DR_f$.

**Solution.** The minimum detectable input and output signals, from (4.7.4) and (4.7.5), assuming that $X = 3$ dB, are

$$P_{i,\text{mds}} = -174 \text{ dBm} + 10 \log (500 \times 10^6) \text{ dB} + 7 \text{ dB} + 3 \text{ dB} = -77 \text{ dBm}$$

and

$$P_{o,\text{mds}} = -77 \text{ dBm} + 40 \text{ dB} = -37 \text{ dBm}$$

Therefore,

$$DR = P_{1\text{dB}} - P_{o,\text{mds}} = 25 \text{ dBm} + 37 \text{ dBm} = 62 \text{ dB}$$

The third-order intercept point, from (4.7.8), is

$$P_{IP} = 25 \text{ dBm} + 10 \text{ dB} = 35 \text{ dBm}$$

and

$$DR_f = \tfrac{2}{3}(\overset{P_{IP}}{25 \text{ dBm}} + \overset{-P_{o,\text{mds}}}{37 \text{ dBm}}) = 48 \text{ dB}$$

Another source of signal distortion is caused by a nonlinear phase characteristic. For a signal to be amplified with no distortion, the magnitude of the power gain transfer function must be constant as a function of frequency and the phase must be a linear function of frequency. A linear phase shift produces a constant time delay to signal frequencies, and a nonlinear phase shift produces different time delays to different frequencies.

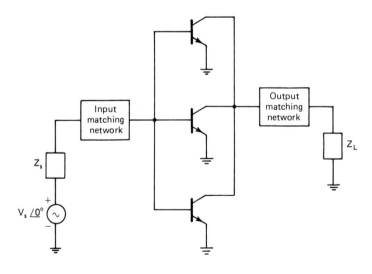

**Figure 4.7.6** Method for paralleling transistors.

## Sec. 4.7 High-Power Amplifier Design

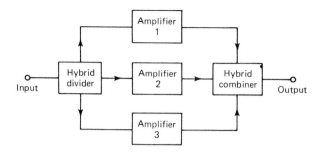

**Figure 4.7.7** A hybrid combiner/divider.

A phase distortion called *AM-to-PM conversion* occurs when an AM signal is transmitted through a power amplifier. The phase shift becomes a function of the instantaneous amplitude of the signal, and the output phase consists of a mean value with a small ripple. The AM-to-PM conversion is defined as the change in output phase for a 1-dB increment of the input power.

Power transistors are provided with flanges or studs for proper mounting and heat dissipation. The maximum junction temperature for a BJT is around 200°C and the maximum channel temperature for a GaAs FET is around 175°C.

When more power is required than can be provided by a single microwave transistor amplifier, power-combining techniques are used. One can use a method of paralleling several transistors, as shown in Fig. 4.7.6. However, this method is not recommended, for several reasons:

1. The input and output impedance levels can be of the same order as the losses in the input and output matching networks. For example, a 0.1-nH inductor having a $Q$ of 150, at $f = 400$ MHz, has a resistance loss of $R = \omega L/Q = 1.6 \ \Omega$, which can be similar to the input resistance of sev-

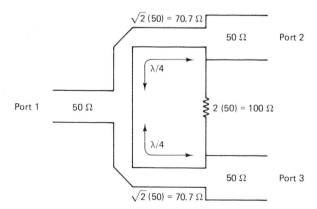

**Figure 4.7.8** The Wilkinson coupler.

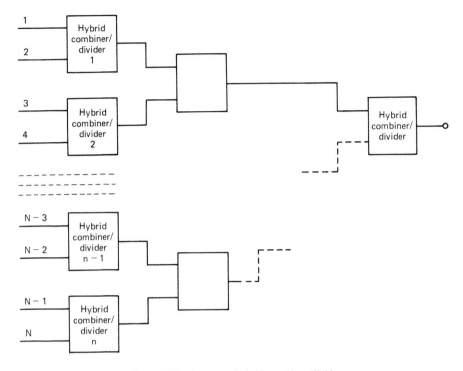

**Figure 4.7.9** An *n*-way hybrid combiner/divider.

eral paralleled transistors. Therefore, the total power output that can be obtained from several paralleled transistors is less than the theoretical total output power because the efficiency decreases as the number of transistors increases.

2. If one transistor fails, the complete amplifier network fails.

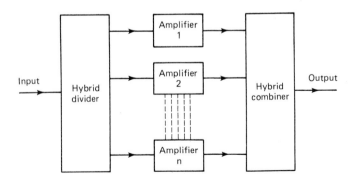

**Figure 4.7.10** An *n*-way power amplifier.

## Sec. 4.7 High-Power Amplifier Design

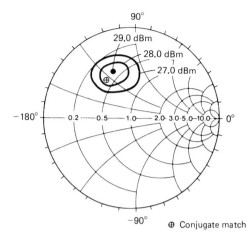

**Figure 4.7.11** Output power contours at 2 GHz, $V_{CE} = 18$ V, and $I_C = 110$ mA. (From Ref. [4.12]; courtesy of Hewlett-Packard.)

3. All transistors must be well matched for power output and gain in order to obtain good load sharing.

A method that avoids the problem of paralleling power transistors is shown in Fig. 4.7.7. It uses a hybrid divider and a hybrid combiner to divide the input power equally to several amplifiers, and to combine the output power of each amplifier. The failure of one amplifier does not cause failure of the complete unit. The complete unit will continue to operate with reduced output power.

A popular two-way hybrid divider known as the *Wilkinson coupler* is shown in Fig. 4.7.8. It consists of two $\lambda/4$ transmission lines with characteristic impedances of $Z_o = \sqrt{2}\,(50) = 70.7\ \Omega$. The input signal is connected to port 1 and divides equally, both in amplitude and phase, when ports 2 and 3 are equally terminated. No power is dissipated in the 100-$\Omega$ resistor when equal loads are connected to ports 2 and 3. If ports 2 and 3 are terminated in 50 $\Omega$, the input impedance of port 1 is the parallel combination of the two 50-$\Omega$ loads, after each is transformed by the $\lambda/4$ line with $Z_o = 70.7\ \Omega$. That is, each 50-$\Omega$ load transforms to 100 $\Omega$, and port 1 sees a 50-$\Omega$ matched input impedance. When a mismatch occurs at port 2 or 3, the reflected signals split through the two transmission lines, travel to the input port, split again, and travel back to the output ports. That is, the reflected wave returns to the output port in two parts, each 180° out of phase from each other. The value of the resistor $2Z_o = 100\ \Omega$ was selected so that the two parts of the reflected wave have equal amplitude and, therefore, perfect cancellation results.

Of course, the two-way hybrid divider in Fig. 4.7.8 can be used as a two-way combiner by applying the input signals at ports 2 and 3 and taking the output at port 1.

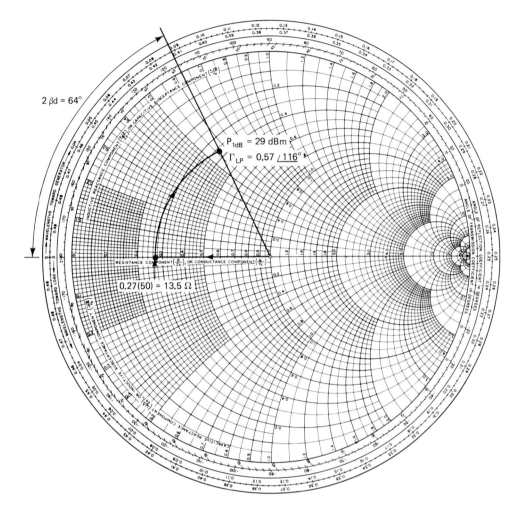

**Figure 4.7.12** Design of the output matching network.

Figure 4.7.9 shows the block diagram of an *n*-way power combiner/divider. The insertion losses of the coupler limit the overall efficiency. The block diagram of an *n*-way amplifier is shown in Fig. 4.7.10.

### Example 4.7.2

Design a power amplifier at 2 GHz using a BJT. The $S$ parameters of the transistor and power characteristics at 2 GHz are

$$S_{11} = 0.64 \underline{|153°}$$
$$S_{21} = 2.32 \underline{|10°}$$

Sec. 4.7    High-Power Amplifier Design                                           185

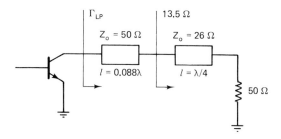

**Figure 4.7.13**  Output network for the 2-GHz amplifier.

$$S_{12} = 0.07 \underline{/-8°}$$

$$S_{22} = 0.51 \underline{/-119°}$$

$$P_{1dB} = 29 \text{ dBm}$$

$$G_{1dB} = 11.5 \text{ dB}$$

Output power contours are shown in Fig. 4.7.11. This figure shows the loci of equal $P_{1dB}$ for different output loading. The input was conjugately matched at all times. The $P_{1dB}$ point and the output conjugate match point were close. (This example is based on a design from Hewlett-Packard Application Note 972 [4.12].)

**Solution.** The transistor is unconditionally stable at 2 GHz since $K = 1.15$ and $\Delta = 0.207 \underline{/58.5°}$. The output network is designed to provide the output power $P_{1dB} = 29$ dBm. The output matching network design is shown in Fig. 4.7.12. The 50-$\Omega$ load was transformed to a resistance of 13.5 $\Omega$ using a quarter-wave transformer with characteristic impedance $Z_o$ given by

$$Z_o = \sqrt{50(13.5)} = 26 \text{ }\Omega$$

A transmission line of length $\beta d = 32°$ (i.e., $0.088\lambda$) was used to complete the match. The output network schematic is shown in Fig. 4.7.13.

In order to obtain an output power of 29 dBm, the input must be conjugately matched. The input conjugate match is calculated using (3.2.5), namely

$$\Gamma_{SP} = (\Gamma_{IN})^* = \left[ 0.64 \underline{/153°} + \frac{(0.07 \underline{/-8°})(2.32 \underline{/10°})(0.57 \underline{/116°})}{1 - (0.51 \underline{/-119°})(0.57 \underline{/116°})} \right]^*$$

$$= 0.749 \underline{/-147.1°}$$

The small-signal $S$ parameters were used to calculate $\Gamma_s$ since, for this transistor, at $P_{1dB}$ the behavior can be assumed to be linear.

The design of the input matching network is illustrated in Fig. 4.7.14. An open-circuited shunt stub of length $0.185\lambda$ ($Z_o = 50$ $\Omega$), followed by a 50-$\Omega$ series transmission line of length 3.5° ($0.0097\lambda$), were used to obtain the match. The complete ac schematic is shown in Fig. 4.7.15.

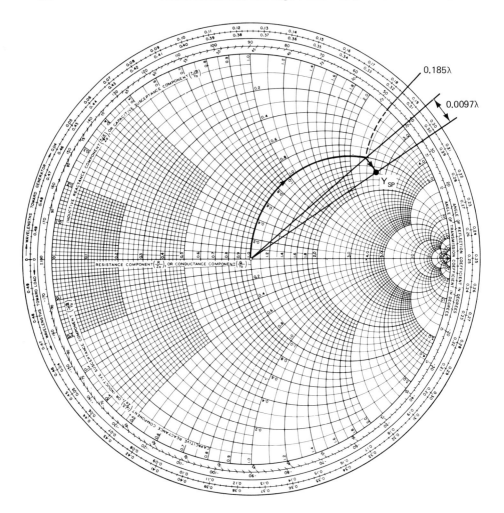

**Figure 4.7.14** Design of the input matching network.

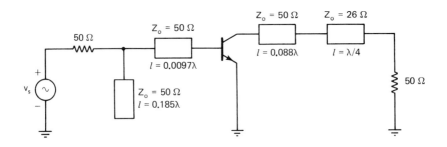

**Figure 4.7.15** Schematic of the power amplifier.

## 4.8 TWO-STAGE AMPLIFIER DESIGN

The configuration of a two-stage microwave transistor amplifier is shown in Fig. 4.8.1. The design of a two-stage amplifier usually consists in the optimization of one of the following requirements: (1) overall high gain, (2) overall low noise figure, or (3) overall high power. In a two-stage amplifier the stability of the individual stages, as well as the overall stability, must be checked.

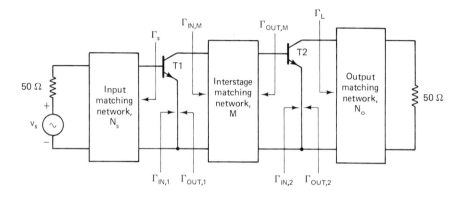

**Figure 4.8.1** Diagram of a two-stage amplifier.

In a design requiring overall high gain, the reflection coefficients are selected as follows:

$$\Gamma_s = (\Gamma_{IN,1})^*$$

$$\Gamma_{IN,M} = (\Gamma_{OUT,1})^*$$

$$\Gamma_{OUT,M} = (\Gamma_{IN,2})^*$$

$$\Gamma_L = (\Gamma_{OUT,2})^*$$

In a design requiring high power the reflection coefficients are selected as follows:

$$\Gamma_s = (\Gamma_{IN,1})^*$$

$$\Gamma_{IN,M} = \Gamma_{LP,1}$$

$$\Gamma_{OUT,M} = (\Gamma_{IN,2})^*$$

$$\Gamma_L = \Gamma_{LP,2}$$

where $\Gamma_{LP,1}$ and $\Gamma_{LP,2}$ are the large-signal load reflection coefficients of $T1$ and $T2$. In other words, the design of $N_o$ results in $\Gamma_L = \Gamma_{LP,2}$ and the design of $M$ is for a conjugate match at its output [i.e., $\Gamma_{OUT,M} = (\Gamma_{IN,2})^*$] and at its input $\Gamma_{IN,M} = \Gamma_{LP,1}$ (i.e., to present $\Gamma_{LP,1}$ to transistor $T1$). The network $N_s$ presents a conjugate match at the input of transistor $T1$.

In a low-noise design, the reflection coefficients are selected as follows:

$$\Gamma_s = \Gamma_{o,1}$$
$$\Gamma_{IN,M} = (\Gamma_{OUT,1})^*$$
$$\Gamma_{OUT,M} = \Gamma_{o,2}$$
$$\Gamma_L = (\Gamma_{OUT,2})^*$$

where $\Gamma_{o,1}$ and $\Gamma_{o,2}$ are the optimum noise source reflection coefficients for stages 1 and 2, respectively.

From (4.2.4), the overall noise figure of a two-stage amplifier depends on $F_1$, $F_2$, and $G_{A1}$. The transistor of the first stage is usually selected to have low noise figure and a higher noise figure is permitted in the second stage. Although some trade-offs between noise figure and gain are possible, usually the optimum noise match with $\Gamma_{o,1}$ and $\Gamma_{o,2}$ is used.

## PROBLEMS

**4.1.** (a) Show that the noise figure for a three-stage amplifier is given by

$$F = F_1 + \frac{F_2 - 1}{G_{A1}} + \frac{F_3 - 1}{G_{A1} G_{A2}}$$

where $F_1$, $F_2$, and $F_3$ are the noise figures of the first, second, and third stages; and $G_{A1}$ and $G_{A2}$ are the available power gains of the first and second stages.

(b) Two cascade amplifiers have noise figures of $F_1 = 1$ dB and $F_2 = 3$ dB, and a gain of $G_{A1} = 10$ dB and $G_{A2} = 16$ dB. Calculate the overall noise figure.

**4.2.** The scattering and noise parameters of a GaAs FET measured at a low-noise bias point ($V_{DS} = 5$ V, $I_D = 15\% I_{DSS} = 10$ mA) at $f = 12$ GHz are

$S_{11} = 0.75 \underline{\vert -116°}$     $F_{min} = 2.2$ dB

$S_{12} = 0.01 \underline{\vert 67°}$     $\Gamma_o = 0.65 \underline{\vert 120°}$

$S_{21} = 3.5 \underline{\vert 64°}$     $R_N = 10 \; \Omega$

$S_{22} = 0.77 \underline{\vert -65°}$

Design a microwave transistor amplifier to have a minimum noise figure.

**4.3.** The scattering and noise parameters of a GaAs FET measured at a low-noise bias point ($V_{DS} = 3.5$ V, $I_D = 15\% I_{DSS} = 12$ mA) at $f = 2$ GHz are

$S_{11} = 0.8 \underline{\vert -51.9°}$     $F_{min} = 1.25$ dB

$S_{12} = 0.045 \underline{\vert 54.6°}$     $\Gamma_o = 0.73 \underline{\vert 60°}$

$S_{21} = 2.15 \underline{\vert 128.3°}$     $R_N = 19.4 \; \Omega$

$S_{22} = 0.73 \underline{\vert -30.5°}$

Design a microwave transistor amplifier to have a minimum noise figure.

**4.4.** (a) A manufacturer provides the information shown in Fig. P4.4 for two microwave transistors. Evaluate the noise parameters for each transistor. Are the constant-gain circles given for $G_T$, $G_A$, or $G_p$?
(b) Verify the location of the $G_A = 10.7$ dB constant-gain circle in Fig. 4.3.4.

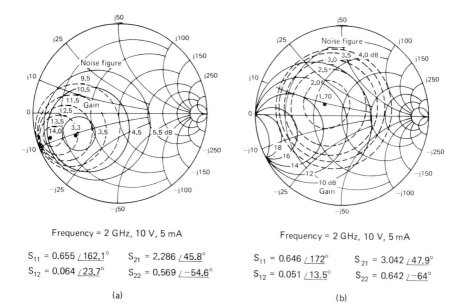

Frequency = 2 GHz, 10 V, 5 mA

$S_{11} = 0.655 \underline{/162.1°}$   $S_{21} = 2.286 \underline{/45.8°}$
$S_{12} = 0.064 \underline{/23.7°}$   $S_{22} = 0.569 \underline{/-54.6°}$

(a)

Frequency = 2 GHz, 10 V, 5 mA

$S_{11} = 0.646 \underline{/172°}$   $S_{21} = 3.042 \underline{/47.9°}$
$S_{12} = 0.051 \underline{/13.5°}$   $S_{22} = 0.642 \underline{/-64°}$

(b)

**Figure P4.4** Two microwave transistors. (From Avantek Transistor Designers Catalog; courtesy of Avantek.)

**4.5.** (a) Design a microwave transistor amplifier at 2 GHz to have a minimum noise figure using the transistor in Fig. P4.4a. Specify the resulting $G_T$, $G_A$, and $G_p$.
(b) Design a microwave transistor amplifier at 2 GHz to have $F_i = 2.5$ dB using the transistor in Fig. P4.4b. Specify the resulting $G_T$, $G_A$, and $G_p$.

**4.6.** (a) In the network shown in Fig. P4.6, verify that $|\Gamma_a| = |\Gamma_{IN}|$, where

$$|\Gamma_a| = \left|\frac{Z_a - Z_o}{Z_a + Z_o}\right|$$

and

$$|\Gamma_{IN}| = \left|\frac{Z_{IN} - Z_s^*}{Z_{IN} + Z_s}\right|$$

(b) Show that $|\Gamma_a|$ can also be expressed in the form

$$|\Gamma_a| = \left|\frac{\Gamma_{IN} - \Gamma_s^*}{1 - \Gamma_{IN}\Gamma_s}\right|$$

**4.7.** Verify (4.4.1), (4.4.2), and (4.4.3).

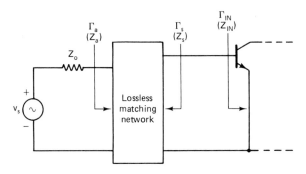

**Figure P4.6**

**4.8.** Design a broadband microwave BJT amplifier to have $G_{TU} = 12$ dB from 150 to 400 MHz. The transistor S parameters are as follows:

| $f$ (MHz) | $S_{11}$ | $S_{21}$ | $S_{22}$ |
|---|---|---|---|
| 150 | $0.31 \underline{|-36°}$ | 5 | $0.91 \underline{|-6°}$ |
| 250 | $0.29 \underline{|-55°}$ | 4 | $0.86 \underline{|-15°}$ |
| 400 | $0.25 \underline{|-76°}$ | 2.82 | $0.81 \underline{|-26°}$ |

(This problem is based on an example given in Ref. [4.13].)

*Hint:* $G_s$ will be small and only $G_L$ should be used to compensate for the variations of $|S_{21}|$ with frequency. $G_s$ should be designed to provide a good VSWR at the input. The output matching network can be designed as in Fig. 4.4.4, or for a better match at the three frequencies an extra inductor in series with the 50-Ω load can be added.

**4.9.** Design a broadband microwave BJT amplifier with $G_{TU} = 10$ dB from 1 to 2 GHz with a noise figure of less than 4 dB. For this transistor $S_{12}$ can be neglected and the scattering and noise parameters are as follows:

| $f$ (GHz) | $S_{11}$ | $S_{21}$ | $S_{22}$ | $\Gamma_o$ | $R_N$ | $F_{min}$ (dB) |
|---|---|---|---|---|---|---|
| 1 | $0.64 \underline{|-98°}$ | $5.04 \underline{|113°}$ | $0.79 \underline{|-30°}$ | $0.48 \underline{|23°}$ | 23.3 | 1.45 |
| 1.5 | $0.60 \underline{|-127°}$ | $3.90 \underline{|87°}$ | $0.76 \underline{|-35°}$ | $0.45 \underline{|61°}$ | 15.6 | 1.49 |
| 2 | $0.59 \underline{|-149°}$ | $3.15 \underline{|71°}$ | $0.75 \underline{|-43°}$ | $0.41 \underline{|88°}$ | 15.7 | 1.61 |

In this problem, design the input and output matching networks so that $G_{TU} = 10$ dB at the band edges only, and calculate the resulting gain at 1.5 GHz. The noise figure over the band must be less than 4 dB.

**4.10.** (a) In the network shown in Fig. P4.10a, what is the best $\Gamma_x$ that can be achieved in the frequency range 400 to 600 MHz?
(b) In the network shown in Fig. P4.10b, what is the best $\Gamma_x$ that can be obtained in the range 6 to 12 GHz?

**4.11.** For the networks in Fig. 4.4.15b, c, and d, the gain–bandwidth restrictions are

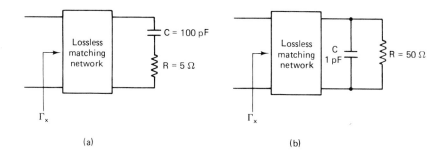

(a)  (b)

Figure P4.10

given by

$$\int_0^\infty \frac{1}{\omega^2} \ln\left|\frac{1}{\Gamma}\right| d\omega \leq \pi RC \quad \text{(for Fig. 4.4.15b)}$$

$$\int_0^\infty \frac{1}{\omega^2} \ln\left|\frac{1}{\Gamma}\right| d\omega \leq \frac{\pi L}{R} \quad \text{(for Fig. 4.4.15c)}$$

$$\int_0^\infty \ln\left|\frac{1}{\Gamma}\right| d\omega \leq \frac{\pi R}{L} \quad \text{(for Fig. 4.4.15d)}$$

Show that $\Gamma_x$ can be expressed in the form given in (4.4.8) when the appropriate definitions of $Q_1$ and $Q_2$ are used. These are given in Fig. 4.4.15.

**4.12.** (a) Verify the equation for $\delta_{IN}$ in (4.5.1).
(b) Verify the relation between $\delta_{IN}$ and $\Gamma_L$ in (4.5.2).
(c) Derive a relation for $\delta_{OUT}$.

**4.13.** (a) Show that $K$ for the balance amplifier, using a 3-dB Lange coupler, is given by

$$K = \frac{1 + P^2}{2P}$$

where

$$P = |S_{21} S_{12}|$$

Show that $K$ has a minimum value of 1 when $P = 1$ (i.e., when $|S_{21} S_{12}| = 1$) and $K > 1$ for all other values of $P$.

(b) The $S$ parameters of a transistor at 2.1 GHz are

$$S_{11} = -0.699 - j0.348$$
$$S_{21} = -10.9 + j7.895$$
$$S_{22} = 0.309 + j0.459$$
$$S_{12} = 0.009 + j0.015$$

and the resulting $K$, $K = 0.48$, shows that the transistor is potentially unstable. If the transistor is used in a balanced amplifier configuration, calculate the resulting $S$ parameters and $K$.

**4.14.** The $S$ parameters of a transistor at 8 GHz are

$$S_{11} = 0.75 \underline{/-100°}$$
$$S_{21} = 2.5 \underline{/93°}$$
$$S_{12} = 0$$
$$S_{22} = 0.7 \underline{/-50°}$$

Determine
(a) The terminations for $G_{TU,\max}$.
(b) The inherent bandwidths and $Q$ of the input and output networks.
(c) The additional elements ($C$ or $L$) that must be added to the input and/or output networks to make the bandwidth 20% of the inherent bandwidth.
(d) The optimum terminations for part (c) and the resulting $G_{TU}$.

**4.15.** In a microwave transistor amplifier it is found that $\Gamma_{Ms} = 0.476 \underline{/166°}$ and $\Gamma_{ML} = 0.846 \underline{/72°}$ at $f = 4$ GHz. Calculate the amplifier input and output intrinsic bandwidths. If the bandwidth is to be limited to 400 MHz, find the value of $C_{OUT}$ or $L_{OUT}$ that must be added to the output network.

**4.16.** Analyze the Wilkinson coupler shown in Fig. 4.7.8.

**4.17.** An amplifier has a transducer power gain of 30 dB, 800-MHz bandwidth, and a noise figure of 5 dB. The 1-dB gain compression point is given as 28 dBm. Calculate DR, $DR_f$, and the maximum output power for no third-order intermodulation distortion.

**4.18.** The specifications for two power GaAs FETs at 4 GHz are as follows:

|       | $P_{1dB}$ (dBm) | $G_{1dB}$ (dB) | $G_p$ (dB) |
|-------|-----------------|----------------|------------|
| FET 1 | 25              | 6              | 7          |
| FET 2 | 20              | 8              | 9          |

Show that the two-way power amplifier shown in Fig. P4.18 can be used to deliver $P_{OUT} = 27.7$ dBm, at 1 dB compression, with $P_{IN} = 6.3$ dBm. The loss of the two-way combiner/divider is $-0.3$ dB. Specify the FET that must be used in each stage, and indicate the power and gain levels at all points.

*Hint:* The power at the output of the divider is 19.0 dBm.

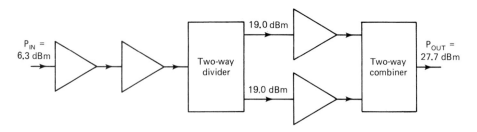

Figure P4.18

**4.19.** Design a power amplifier at 4 GHz using a BJT. The $S$ parameters of the transistor and power characteristics at 4 GHz are

$$S_{11} = 0.32 \underline{/-145°}$$
$$S_{21} = 1.38 \underline{/-113°}$$
$$S_{12} = 0.08 \underline{/-98°}$$
$$S_{22} = 0.8 \underline{/-177°}$$
$$P_{1dB} = 27.5 \text{ dBm}$$
$$G_{1dB} = 7 \text{ dB}$$
$$\Gamma_{LP} = 0.1 \underline{/0°}$$

**4.20.** In the two-stage amplifier shown in Fig. 4.8.1, transistors $T1$ and $T2$ are potentially unstable. Can the overall amplifier be unconditionally stable?

## REFERENCES

[4.1] H. A. Haus (chairman), "Representation of Noise in Linear Two Ports," IRE Subcommittee 7.9 on Noise, *Proceedings of the IEEE*, January 1960.

[4.2] "A Low Noise 4 GHz Transistor Amplifier Using the HXTR-6101 Silicon Bipolar Transistor," Hewlett-Packard Application Note 967, May 1975.

[4.3] "A 6 GHz Amplifier Using the HFET-1101 GaAs FET," Hewlett-Packard Application Note 970, February 1978.

[4.4] T. T. Ha, *Solid State Microwave Amplifier Design*, Wiley-Interscience, New York, 1981.

[4.5] G. Matthaei, L. Young, and E. M. T. Jones, *Microwave Filters, Impedance-Matching Networks, and Coupling Structures*, Artech House, Inc., Dedham, MA., 1980.

[4.6] AMPSYN (computer program), Compact Software, Inc., 1131 San Antonio Road, Palo Alto, CA 94303.

[4.7] D. J. Mellor and J. G. Linvill, "Synthesis of Interstage Networks of Prescribed Gain versus Frequency Slopes," *IEEE Transactions on Microwave Theory and Techniques*, December 1975.

[4.8] R. W. Anderson, "S Parameter Techniques for Faster, More Accurate Network Design," Hewlett-Packard Application Note 95-1 (or *Hewlett-Packard Journal*), February 1967.

[4.9] J. Lange, "Integrated Stripline Quadrature Hybrids," *IEEE Transactions on Microwave Theory and Techniques*, December 1969.

[4.10] L. Besser, "Microwave Circuit Design," *Electronic Engineering*, October 1980.

[4.11] R. M. Fano, "Theoretical Limitations on the Broadband Matching of Arbitrary Impedances," *Journal of the Franklin Institute*, January 1950.

[4.12] "Two Telecommunications Power Amplifiers for 2 and 4 GHz Using the HXTR-5102 Silicon Bipolar Power Transistor," Hewlett-Packard Application Note 972, 1980.

[4.13] R. S. Carson, *High-Frequency Amplifiers*, Wiley-Interscience, New York, 1975.

# 5
# MICROWAVE TRANSISTOR OSCILLATOR DESIGN

## 5.1 INTRODUCTION

In this chapter the analytical techniques that are used in the design of negative-resistance oscillators are discussed. The small- and large-signal $S$ parameters provide all the information needed to design negative-resistance oscillators.

In a negative-resistance oscillator we refer to the matching networks at the two ports as the terminating and the load (or resonant) matching networks. The load-matching network is the network that determines the frequency of oscillation, and the terminating network is used to provide the proper matching.

The design of the terminating and load matching networks must be done carefully. For example, the condition $|\Gamma_{IN}| > 1$ is necessary for oscillation. A short circuit at the terminating port can produce $|\Gamma_{IN}| > 1$. However, no power is delivered to a short-circuited termination.

In the low range of microwave frequencies, the lumped-element Colpitts, Hartley, and Clapp oscillators are commonly used.

## 5.2 ONE-PORT NEGATIVE-RESISTANCE OSCILLATORS

A general schematic diagram for one-port negative-resistance oscillators is shown in Fig. 5.2.1. The negative-resistance device is represented by the amplitude and frequency-dependent impedance

$$Z_{IN}(V, \omega) = R_{IN}(V, \omega) + jX_{IN}(V, \omega)$$

## Sec. 5.2 One-Port Negative-Resistance Oscillators

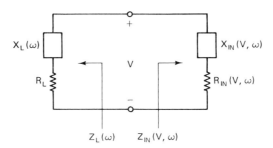

**Figure 5.2.1** Schematic diagram for one-port negative-resistance oscillators.

where

$$R_{IN}(V, \omega) < 0$$

The oscillator is constructed by connecting the device to a passive load impedance, called

$$Z_L(\omega) = R_L + jX_L(\omega)$$

The discussion in Section 3.3 showed that the one-port network in Fig. 5.2.1 is stable if

$$\text{Re}\,[Z_{IN}(V, \omega) + Z_L(\omega)] > 0$$

and the network will oscillate when

$$\Gamma_{IN}(V, \omega)\Gamma_L(\omega) = 1$$

The oscillation conditions can be expressed in the form

$$R_{IN}(V, \omega) + R_L = 0 \qquad (5.2.1)$$

and

$$X_{IN}(V, \omega) + X_L(\omega) = 0$$

To be specific, the device is defined to be unstable over some frequency range $\omega_1 < \omega < \omega_2$ if $R_{IN}(V, \omega) < 0$. The one-port network is unstable for some $\omega_o$ in the range if the net resistance of the network is negative, that is, when

$$|R_{IN}(V, \omega_o)| > R_L \qquad (5.2.2)$$

Any transient excitation due to noise in the circuit will initiate an oscillation at the frequency $\omega_o$, for which the net reactance of the network is equal to zero, namely

$$X_L(\omega_o) = -X_{IN}(V, \omega_o) \qquad (5.2.3)$$

At $\omega_o$ a growing sinusoidal current will flow through the circuit, and the oscillation will continue to build up as long as the resistance is negative. The amplitude of the voltage must eventually reach a steady-state value, called $V_o$, which occurs when the loop resistance is zero. To satisfy the conditions (5.2.1) and (5.2.2), the impedance $Z_{IN}(V, \omega)$ must be amplitude dependent and, therefore, at $V = V_o$ we can write

$$R_{IN}(V_o, \omega_o) + R_L = 0 \qquad (5.2.4)$$

The frequency of oscillation determined by (5.2.3) is not stable since $X_{IN}(V, \omega_o)$ is amplitude dependent. That is,

$$X_{IN}(V_1, \omega_o) \neq X_{IN}(V_o, \omega_o)$$

where $V_1$ is an arbitrary voltage. Therefore, it is necessary to find another condition to guarantee a stable oscillation. If the frequency dependence of $Z_{IN}(V, \omega)$ can be neglected for small variations around $\omega_o$, Kurokawa [5.1] has shown that the condition for a stable oscillation is

$$\left.\frac{\partial R_{IN}(V, \omega)}{\partial V}\right|_{V=V_o} \left.\frac{dX_L(\omega)}{d\omega}\right|_{\omega=\omega_o} - \left.\frac{\partial X_{IN}(V, \omega)}{\partial V}\right|_{V=V_o} \left.\frac{dR_L(\omega)}{d\omega}\right|_{\omega=\omega_o} > 0 \qquad (5.2.5)$$

In other words, the frequency of oscillation determined by (5.2.3) and (5.2.4) is stable only if (5.2.5) is satisfied. In most cases

$$\frac{dR_L(\omega)}{d\omega} = 0$$

(i.e., $R_L$ is a constant) and (5.2.5) simplifies accordingly.

**Example 5.2.1**

A negative-resistance device can be modeled by the parallel combination of a capacitor and a negative conductance, as shown in Fig. 5.2.2a. The amplitude dependence of the negative conductance, shown in Fig. 5.2.2b, is given by

$$G(V) = G_M\left(1 - \frac{V}{V_M}\right) \qquad (5.2.6)$$

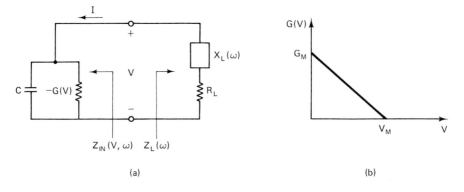

**Figure 5.2.2** (a) Negative-resistance device; (b) amplitude variations of $G(V)$.

## Sec. 5.2 One-Port Negative-Resistance Oscillators

Design a load circuit, $Z_L$, to provide oscillation at $\omega_o$ and calculate the output power.

**Solution.** The device impedance is

$$Z_{IN}(V, \omega) = R_{IN}(V, \omega) + jX_{IN}(V, \omega) = \frac{-G(V)}{G^2(V) + \omega^2 C^2} + j\frac{-\omega C}{G^2(V) + \omega^2 C^2} \quad (5.2.7)$$

From (5.2.3) to (5.2.5) a stable oscillation at $\omega = \omega_o$ occurs when

$$R_L = \frac{G(V)}{G^2(V) + \omega^2 C^2}\bigg|_{\omega=\omega_o, V=V_o} \quad (5.2.8)$$

$$X_L(\omega) = \frac{\omega C}{G^2(V) + \omega^2 C^2}\bigg|_{\omega=\omega_o, V=V_o} \quad (5.2.9)$$

and

$$\frac{\partial R_{IN}}{\partial V}\bigg|_{V=V_o} \frac{dX_L}{d\omega}\bigg|_{\omega=\omega_o} > 0 \quad (5.2.10)$$

where $V_o$ is the oscillation voltage level at the frequency of oscillation $\omega_o$.

Substituting (5.2.6) into (5.2.7) gives for $R_{IN}$,

$$R_{IN} = \frac{-(1/G_M)(1 - V/V_M)}{(1 - V/V_M)^2 + \omega^2 C^2/G_M^2} \quad (5.2.11)$$

Differentiating (5.2.11) with respect to $V$ gives

$$\frac{\partial R_{IN}}{\partial V} = \frac{-1 + 2(V/V_M) - V^2/V_M^2 + \omega^2 C^2/G_M^2}{G_M V_M[(1 - V/V_M)^2 + (\omega^2 C^2/G_M^2)]^2} \quad (5.2.12)$$

and substituting (5.2.12) into (5.2.10) produces the relation

$$\frac{dX_L}{d\omega}\bigg|_{\omega=\omega_o} \left[-1 + 2\frac{V}{V_M} - \frac{V^2}{V_M^2} + \frac{\omega^2 C^2}{G_M^2}\right]\bigg|_{V=V_o} > 0 \quad (5.2.13)$$

There is no direct way to solve for $R_L$ and $X_L$ from (5.2.8), (5.2.9), and (5.2.13). Therefore, another design consideration such as maximizing the power delivered to $R_L$ must be introduced.

The current in the circuit is given by

$$I = V(-G(V) + j\omega C)$$

and the output power is given by

$$P = \tfrac{1}{2}|I|^2 R_L = \tfrac{1}{2}|V|^2 R_L(G^2(V) + \omega^2 C^2) \quad (5.2.14)$$

Substituting (5.2.6) and (5.2.8) into (5.2.14) gives

$$P = \frac{1}{2}V_M^2\left(1 - \frac{G(V)}{G_M}\right)^2 G(V) \quad (5.2.15)$$

The expression (5.2.15) can be maximized for $G$ as follows:

$$\frac{\partial P}{\partial G(V)} = \frac{1}{2}V_M^2\left[1 - 4\frac{G(V)}{G_M} + 3\frac{G^2(V)}{G_M^2}\right] = 0$$

or

$$\frac{G(V)}{G_M} = \frac{1}{3} \tag{5.2.16}$$

Substituting (5.2.16) into (5.2.6) gives

$$\frac{V}{V_M} = \frac{2}{3} \tag{5.2.17}$$

which is the output voltage when maximum output power is delivered to $R_L$. If (5.2.17) is evaluated at $V = V_o$ and substituted back into (5.2.13), the following result is obtained:

$$\left.\frac{dX_L}{d\omega}\right|_{\omega=\omega_o} \left(\frac{\omega_o^2 C^2}{G_M^2} - \frac{1}{9}\right) > 0 \tag{5.2.18}$$

The only unknown in (5.2.18) is $X_L$, so the frequency dependence of $X_L$ can easily be determined around $\omega_o$.

From (5.2.16), (5.2.8), and (5.2.9) the values of $R_L$ and $X_L(\omega)$ that maximize the power delivered to $R_L$ at $\omega_o$ are

$$R_L = \frac{G_M/3}{(G_M/3)^2 + \omega_o^2 C^2} \tag{5.2.19}$$

and

$$X_L(\omega_o) = \frac{\omega_o C}{(G_M/3)^2 + \omega_o^2 C^2} \tag{5.2.20}$$

At this point, it is necessary to check if $R_L$ satisfies the condition (5.2.2) when the amplitude level is zero (i.e., the starting oscillation condition). Therefore, using (5.2.11) and (5.2.19), we have that

$$|R_{IN}(V, \omega_o)|\Big|_{V=0} > R_L$$

when

$$\frac{\omega_o C}{G_M} > \frac{1}{\sqrt{3}}$$

If we examine the ratio of $R_L$ to $|R_{IN}(0, \omega_o)|$, we obtain

$$\frac{R_L}{|R_{IN}(0, \omega_o)|} = 3 \frac{1 + (\omega_o C/G_M)^2}{1 + 9(\omega_o C/G_M)^2} \tag{5.2.21}$$

If $\omega_o C/G_M$ is large, (5.2.21) can be approximated by

$$\frac{R_L}{|R_{IN}(0, \omega_o)|} \approx \frac{1}{3} \tag{5.2.22}$$

The relation (5.2.22) provides a good design guideline for selecting $R_L$. That is, let

$$R_L = \tfrac{1}{3}|R_{IN}(0, \omega_o)| \tag{5.2.23}$$

Sec. 5.3    Two-Port Negative-Resistance Oscillators    199

From (5.2.20), the frequency of oscillation $\omega_o$ is

$$X_L(\omega_o) \approx \frac{1}{\omega_o C} \quad (5.2.24)$$

and from (5.2.18),

$$\left.\frac{dX_L}{d\omega}\right|_{\omega=\omega_o} > 0 \quad (5.2.25)$$

Obviously, an inductor ($X_L = \omega L$) satisfies (5.2.24) and (5.2.25), and $\omega_o$ is given by

$$\omega_o \approx \frac{1}{\sqrt{LC}}$$

## 5.3 TWO-PORT NEGATIVE-RESISTANCE OSCILLATORS

A two-port network configuration is shown in Fig. 5.3.1. The two-port network is characterized by the $S$ parameters of the transistor, the terminating impedance $Z_T$, and the load impedance $Z_L$.

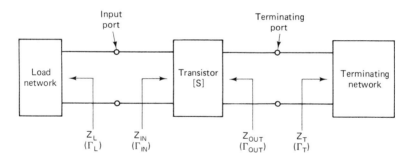

**Figure 5.3.1** Two-port oscillator model.

When the two-port is potentially unstable, an appropriate $Z_T$ permits the two-port to be represented as a one-port negative-resistance device with input impedance $Z_{IN}$, as shown in Fig. 5.2.1. The conditions for a stable oscillation are given by (5.2.3) to (5.2.5).

The negative resistance of $Z_{IN}$ is a function of voltage and as the oscillation power increases, the negative resistance can decrease to a value lower than the load resistance, at which point the oscillation stops. This problem is eliminated by designing the magnitude of the negative resistance, at $V = 0$, to be larger than the load. The value given in (5.2.23) [i.e., $|R_{IN}(0, \omega_o)| = 3R_L$] is commonly used in practice.

When the input port is made to oscillate, the terminating port also oscillates. The fact that both ports are oscillating can be proved as follows.

The input port is oscillating when
$$\Gamma_{IN}\Gamma_L = 1 \tag{5.3.1}$$
and from (3.2.5) and (5.3.1),
$$\Gamma_L = \frac{1}{\Gamma_{IN}} = \frac{1 - S_{22}\Gamma_T}{S_{11} - \Delta\Gamma_T}$$
or
$$\Gamma_T = \frac{1 - S_{11}\Gamma_L}{S_{22} - \Delta\Gamma_L} \tag{5.3.2}$$

Also, from (3.2.6),
$$\Gamma_{OUT} = \frac{S_{22} - \Delta\Gamma_L}{1 - S_{11}\Gamma_L} \tag{5.3.3}$$
and from (5.3.2) and (5.3.3) it follows that
$$\Gamma_{OUT}\Gamma_T = 1$$
which shows that the terminating port is also oscillating.

A design procedure for a two-port oscillator is as follows:

1. Use a potentially unstable transistor at the frequency of oscillation $\omega_o$.
2. Design the terminating network to make $|\Gamma_{IN}| > 1$. Series or shunt feedback can be used to increase $|\Gamma_{IN}|$.
3. Design the load network to resonate $Z_{IN}$. That is, let
$$X_L(\omega_o) = -X_{IN}(\omega_o) \tag{5.3.4}$$
and
$$R_L = \frac{|R_{IN}(0, \omega_o)|}{3} \tag{5.3.5}$$

This design procedure is popular due to its high rate of success. However, the frequency of oscillation will shift somewhat from its designed value at $\omega_o$. This occurs because the oscillation power increases until the negative resistance is equal to the load resistance and $X_{IN}$ varies as a function of $V$ (i.e., as a function of the oscillation power). Also, there is no assurance that the oscillator is providing optimum power.

**Example 5.3.1**

Design an 8-GHz GaAs FET oscillator using the reverse-channel configuration shown in Fig. 5.5.4. The S parameters of the transistor, in the reverse-channel configuration, at 8 GHz are
$$S_{11} = 0.98 \underline{|163°}$$
$$S_{21} = 0.675 \underline{|-161°}$$

## Sec. 5.3  Two-Port Negative-Resistance Oscillators

$$S_{12} = 0.39 \underline{/-54°}$$
$$S_{22} = 0.465 \underline{/120°}$$

(This example is based on a design from Refs. [5.2] and [5.3].)

**Solution.** The transistor is potentially unstable at 8 GHz (i.e., $K = 0.529$) and the stability circle at the gate-to-drain port is shown in Fig. 5.3.2. In the notation of Fig. 5.3.1, the gate-to-drain port is the terminating port.

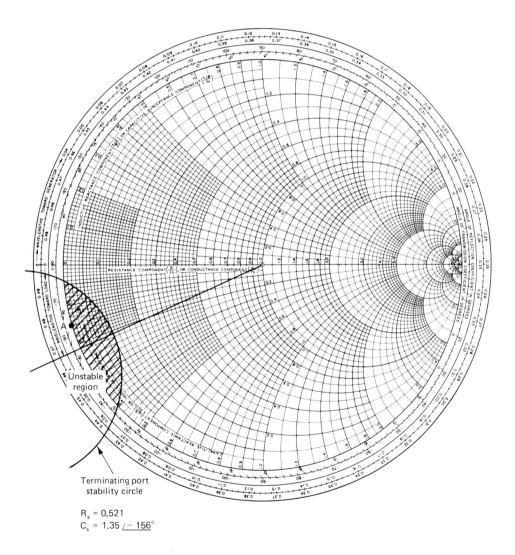

**Figure 5.3.2** Terminating port stability circle.

As shown in Fig. 5.3.2, any $\Gamma_T$ in the shaded region produces $|\Gamma_{IN}| > 1$ (i.e., a negative resistance at the input port). Selecting $\Gamma_T$ at point $A$ in Fig. 5.3.2 (i.e., $\Gamma_T = 1\underline{|-163°}$), the associated impedance is $Z_T = -j7.5\,\Omega$. This reactance can be implemented by an open-circuited 50-$\Omega$ line of length $0.226\lambda$. With $Z_T$ connected, the input reflection coefficient is found to be $\Gamma_{IN} = 12.8\underline{|-16.6°}$, and the associated impedance is $Z_{IN} = -58 - j2.6\,\Omega$. The load matching network is designed using (5.3.4) and (5.3.5), that is, $Z_L = 19 + j2.6\,\Omega$ at $f_o = 8$ GHz.

As reported in Refs. [5.2] and [5.3], the oscillator was constructed and oscillated readily at frequencies between 7.5 and 7.8 GHz, with output power between 680 and 940 mW at $V_{DS} = 9$ V. Some tuning was necessary to move the oscillation frequency to 8 GHz.

**Example 5.3.2**

Design a 2.75-GHz oscillator using a BJT in a common-base configuration. The transistor $S$ parameters at 2.75 GHz are

$$S_{11} = 0.9\underline{|150°}$$
$$S_{21} = 1.7\underline{|-80°}$$
$$S_{12} = 0.07\underline{|120°}$$
$$S_{22} = 1.08\underline{|-56°}$$

(This example is based on a design from Ref. [5.4].)

**Solution.** The transistor is potentially unstable at 2.75 GHz ($K = -0.64$). The instability of the transistor can be increased using external feedback. For the common-base configuration (and also for the common-gate configuration) a common lead inductance from base to ground (as shown in Fig. 5.3.3) is commonly used.

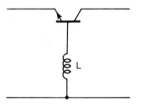

Figure 5.3.3 BJT with external feedback to increase instability.

Varying $L$ from 0.5 nH to 15 nH shows that the instability at the input and output are optimized with $L = 1.45$ nH. With $L = 1.45$ nH, the resulting $S$ parameters for the network in Fig. 5.3.3 are

$$S_{11} = 1.72\underline{|100°}$$
$$S_{21} = 2.08\underline{|-136°}$$
$$S_{12} = 0.712\underline{|94°}$$
$$S_{22} = 1.16\underline{|-102°}$$

and $K = -0.56$.

Sec. 5.4   Oscillator Design Using Large-Signal Measurement

The common lead inductance has been used to raise $|S_{11}|$ and $|S_{22}|$ to large values. Since $|S_{11}| > |S_{22}|$, it appears that the emitter-to-ground port is the best place for the load network (i.e., the tuning network). Of course, these values are obtained with 50-$\Omega$ terminations, and 50-$\Omega$ terminations are not necessarily used for the matching networks.

The terminating network can be designed to present an impedance to the collector having a real part smaller than 50 $\Omega$, and to couple the oscillator to a 50-$\Omega$ termination. A design for the terminating network is illustrated in Fig. 5.3.4. With the values shown in Fig. 5.3.4, $\Gamma_{IN} = 2.21 \underline{/119°}$ ($Z_{IN} = -24 + j24.2\ \Omega$). From (5.3.4) and (5.3.5) the impedance of the load matching network should be $Z_L = 8 - j24.2\ \Omega$.

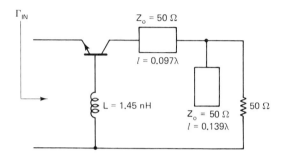

**Figure 5.3.4**   Terminating network design.

## 5.4 OSCILLATOR DESIGN USING LARGE-SIGNAL MEASUREMENTS

In this section a two-step method based on large-signal measurements is developed for oscillator design [5.5]. Basically, the method consists in designing the terminating network so that the two-port presents a large (i.e., optimum) negative resistance at the input port. The resulting one-port negative-resistance network can be placed in a nonoscillating circuit and the optimum load impedance as a function of power (i.e., large-signal measurements) can be measured.

The reflection coefficient $\Gamma_{IN}$ for the network in Fig. 5.3.1 is given by

$$\Gamma_{IN} = \frac{S_{11} - \Delta\Gamma_T}{1 - S_{22}\Gamma_T}$$

which can be manipulated into the form

$$\Gamma_{IN} = \frac{S_{11} - \Delta S_{22}^*}{1 - |S_{22}|^2} + \frac{S_{12}S_{21}}{1 - |S_{22}|^2}\frac{\Gamma_T - S_{22}^*}{1 - S_{22}\Gamma_T}$$

$$= \Gamma_{IN,o} + \alpha\Gamma_T' \tag{5.4.1}$$

where

$$\Gamma_{IN,o} = \frac{S_{11} - \Delta S_{22}^*}{1 - |S_{22}|^2} \quad (5.4.2)$$

$$\alpha = \frac{S_{12} S_{21}}{1 - |S_{22}|^2} \frac{1 - S_{22}^*}{1 - S_{22}} \quad (5.4.3)$$

and

$$\Gamma_T' = \frac{Z_T - Z_{22}^*}{Z_T + Z_{22}} \quad (5.4.4)$$

$Z_{22}$ is the impedance associated with $S_{22}$.

A simple graphical method can be developed to relate $\Gamma_T$ to $\Gamma_{IN}$. The transformation in (5.4.1) shows that the magnitude of $\Gamma_T'$ (i.e., the $\Gamma_T$ plane normalized to $Z_{22}$) is multiplied by $|\alpha|$ and the phase of $\Gamma_T'$ is rotated by $\underline{/\alpha}$. Since $\Gamma_{IN,o}$ is a constant in the $\Gamma_{IN}$ plane, its contribution is to shift the center of $\Gamma_T'$. A typical transformation is illustrated in Fig. 5.4.1. Any $\Gamma_T'$ in the shaded area will cause oscillations.

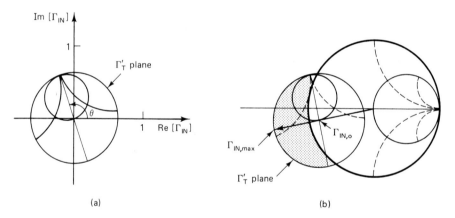

**Figure 5.4.1** (a) $\Gamma_T'$ plane scaled by $\alpha$, where $\theta = \arg(\alpha)$ ; (b) typical mapping from $\Gamma_T'$ plane to $\Gamma_{IN}$ according to (5.4.1).

From (5.4.1), $|\Gamma_{IN}|$ is a maximum when $|\Gamma_T'| = 1$ and $\alpha \Gamma_T'$ is in the direction of $\Gamma_{IN,o}$. That is,

$$\Gamma_{IN,max} = \Gamma_{IN,o} + |\alpha| \hat{u}_{IN,o} \quad (5.4.5)$$

where $\hat{u}_{IN,o}$ is a unit vector in the direction of $\Gamma_{IN,o}$. The value of $\Gamma_{IN,max}$ is illustrated in Fig. 5.4.1b.

The value of $\Gamma_T$ that maximizes $\Gamma_{IN}$, called $\Gamma_{T,o}$, is given by

$$\Gamma_{T,o} = \frac{1 + (\hat{u}_{12}/\hat{u}_{IN,o}) S_{22}^*}{(\hat{u}_{12}/\hat{u}_{IN,o}) + S_{22}} \quad (5.4.6)$$

where $\hat{u}_{12}$ is a unit vector in the direction of $S_{12} S_{21}$. The associated input and terminating impedances are

$$Z_{IN,max} = \frac{1 + \Gamma_{IN,max}}{1 - \Gamma_{IN,max}} \quad (5.4.7)$$

and

$$Z_{T,o} = \frac{1 + \Gamma_{T,o}}{1 - \Gamma_{T,o}} \quad (5.4.8)$$

The value of $\Gamma_{T,o}$ in (5.4.6) produces an optimum $\Gamma_{IN}$ (i.e., $\Gamma_{IN,max}$). In other words, the two-port network has been reduced to an optimum negative-resistance one-port by maximizing the small-signal input reflection coefficient of the transistor. Thus far, only the small-signal $S$ parameters of the transistor were used since the optimized $\Gamma_{IN}$ is the amplitude-independent small-signal input reflection coefficient.

The one-port negative-resistance oscillator can now be characterized by measuring the input impedance as a function of input power at the frequency $\omega_o$. This is a large-signal characterization of the one-port which is also called a *device-line characterization*.

It is a good idea to place $\Gamma_{IN,max}$ within the range shown in Fig. 5.4.2. In this range the associated $|R_{IN,max}|$ is less than 50 $\Omega$ and $X_{IN,max}$ is small. The reason for this selection is that in the design procedure that follows we need to take some measurements at the input port with a 50-$\Omega$ source impedance. Also, the larger the ratio $|R_{IN,max}|/X_{IN,max}$, the larger the $Q$.

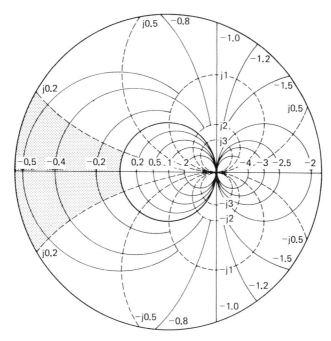

**Figure 5.4.2** Best range for $\Gamma_{IN,max}$.

The method described establishes the terminating impedance that reduces the two-port to an optimum negative-resistance one-port. The one-port can now be characterized by large-signal measurements.

The large-signal characterization is achieved by measuring, in the circuit shown in Fig. 5.4.3, the current $I_D$ and the impedance $Z_{IN}(I_D, \omega_o)$, as $V_s$ is varied. The measurements are made at the desired frequency of oscillation $\omega_o$, and the source resistance is typically 50 Ω. With $|R_{IN,max}|$ selected in the range shown in Fig. 5.4.2, the circuit in Fig. 5.4.3 is stable.

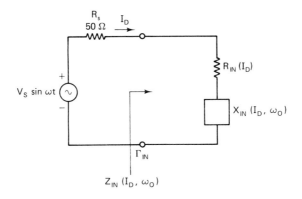

**Figure 5.4.3** Large-signal measuring circuit.

In the circuit shown in Fig. 5.4.3, the current $I_D$ is given by

$$I_D = \frac{V_s}{R_s + R_{IN}(I_D) + jX_{IN}(I_D, \omega_o)} \qquad (5.4.9)$$

and the power delivered by the negative resistance $R_{IN}(I_D)$ is

$$P_D(\omega_o) = \tfrac{1}{2}|I_D|^2 |R_{IN}(I_D)|$$

The measurement of $P_D$ versus $Z_{IN}(I_D, \omega_o)$ generates the large-signal characteristics of the one-port network. If the one-port is now terminated in the load impedance (see Fig. 5.2.1),

$$Z_L(\omega_o) = -Z_{IN}(I_D, \omega_o)$$

the power delivered to $Z_L$ is given by $P_L(\omega_o) = P_D(\omega_o)$.

Obviously, the measurement of $I_D$ at microwave frequencies is difficult. Therefore, in practice the reflection coefficient $\Gamma_{IN}$ as a function of the available power from the source is measured. The available power from the source

### Sec. 5.4  Oscillator Design Using Large-Signal Measurement

is given by
$$P_{AVS} = \frac{V_s^2}{8R_s}$$

The power added $P_{ADD}$ (i.e., the reflected minus the available input power) is given by
$$P_{ADD} = P_{AVS}(|\Gamma_{IN}|^2 - 1)$$
and can be expressed in the form
$$P_{ADD} = \frac{V_s^2 |R_{IN}|}{2[(R_{IN} + R_s)^2 + X_{IN}^2]} \quad (5.4.10)$$

Substituting (5.4.9) into (5.4.10) gives
$$P_{ADD} = \frac{|I_D|^2}{2} |R_{IN}|$$

which shows that the added power is the power that the one-port will deliver to the load $Z_L(\omega_o) = -Z_{IN}(I_D, \omega_o)$.

The large-signal characterizations of the one-port are generated by measuring $\Gamma_{IN}$ and $P_{AVS}$, and calculating $P_{ADD}$ versus $Z_{IN}$ as a function of $I_D$ at the desired frequency of oscillation $\omega_o$.

There are several ways of implementing the load impedance $Z_L(\omega_o)$ and, of course, not all of them will give us a stable oscillation. For a stable oscillation we have to check that $Z_L(\omega_o)$ satisfies the condition given in (5.2.5). This can be achieved easily since the amplitude dependence of $R_{IN}(I_D)$ can be obtained from the measured data, and the required impedance variation (i.e., $dX_L/d\omega \gtrless 0$) can be determined.

In conclusion, the design procedure uses the small-signal $S$ parameters to establish the terminating impedance that results in an optimum negative-resistance one-port network. Then the one-port oscillator performance is described by the measured large-signal characteristics.

### Example 5.4.1

Design an oscillator using a GaAs FET whose $S$ parameters at 10 GHz are
$$S_{11} = 0.9 \underline{|180°}$$
$$S_{12} = 0.79 \underline{|-98°}$$
$$S_{21} = 0.89 \underline{|-163°}$$
$$S_{22} = 0.2 \underline{|180°}$$

(This example is based on a design from Ref. [5.5].)

**Solution.** For this transistor $K = 0.51$, showing that the device is potentially unstable. $\Gamma_{T,o}$ and $\Gamma_{IN,max}$ and the associated $Z_{T,o}$ and $Z_{IN,max}$ can be determined from (5.4.2) to

(5.4.8), namely

$$\Gamma_{IN,o} = 0.89 \underline{|-171°}$$
$$\alpha = 0.76 \underline{|\,99°}$$
$$\Gamma_{T,o} = 1 \underline{|\,114°}$$
$$Z_{T,o} = j0.66$$
$$\Gamma_{IN,max} = 1.65 \underline{|-171°}$$
$$Z_{IN,max} = -0.247 - j0.074$$

The complete mapping of the $\Gamma_T$ plane into the $\Gamma_{IN}$ plane is illustrated in Fig. 5.4.4. Observe the locations of $\Gamma_{IN,o}$ and $\Gamma_{IN,max}$.

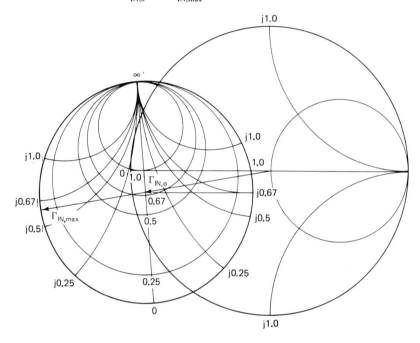

**Figure 5.4.4** Mapping of $\Gamma_T$ plane into the $\Gamma_{IN}$ plane. (From W. Wagner [5.5]; reproduced with permission of *Microwave Journal*.)

With the termination $Z_{T,o} = j0.66$, the transistor large-signal characteristics are now measured. The results from the large-signal measurements are given in Fig. 5.4.5. It is observed that the maximum power added is 45 mW when $Z_L = 4 - j7.5 \, \Omega$.

## 5.5 OSCILLATOR CONFIGURATIONS

At the low end of the microwave frequency range, lumped-element oscillators are commonly used. Three basic oscillator configurations used are the Colpitts, Hartley, and Clapp oscillators. They are shown in Fig. 5.5.1 in a

## Sec. 5.5 Oscillator Configurations

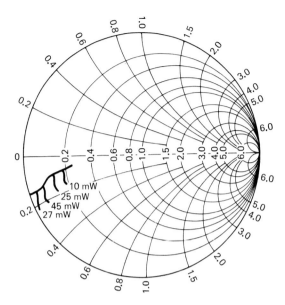

**Figure 5.4.5** Large-signal characteristic at 10 GHz with $Z_{T,o} = j0.66$. (From W. Wagner [5.5]; reproduced with permission of *Microwave Journal*.)

common-base transistor configuration. The Colpitts network uses a capacitor voltage divider in the tuned circuit to provide the correct feedback. The Hartley network uses a tapped inductor tuned circuit, and the Clapp network is similar to the Colpitts network but with an extra capacitor in series with the inductor to improve the frequency stability.

The high-$Q$ tapped inductor required in the Hartley's oscillator is difficult to build. Therefore, the Colpitts and Clapp oscillators are usually preferred.

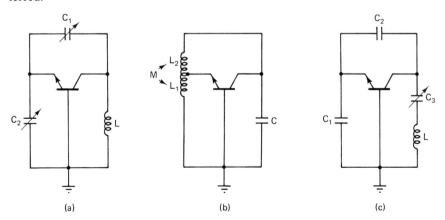

**Figure 5.5.1** Three types of common-base transistor configurations: (a) Colpitts; (b) Hartley; (c) Clapp.

At higher microwave frequencies (i.e., in the gigahertz range), the parasitic capacitances of the packaged transistors provide some or all the feedback needed for oscillation. In this range the negative-resistance design procedure is used, since the $S$ parameters provide all the needed design information. The negative-resistance design procedure basically consists of selecting a transistor in an oscillator topology that provides the required output power. The transistor in the configuration selected must be potentially unstable at the desired frequency of oscillation. Feedback can be added to increase the negative resistance associated with $\Gamma_{IN}$ or $\Gamma_{OUT}$. The terminating and load matching networks must be designed to provide the proper resonance conditions.

For a BJT negative-resistance oscillator the most effective network topology is the common-base configuration. This configuration, illustrated in Fig. 5.5.2, is used in low-power oscillator circuits, and it is easy to tune. The inductor feedback element is used to increase $|\Gamma_{IN}|$ and $|\Gamma_{OUT}|$. Common-emitter and common-collector configurations have also been used in microwave oscillators.

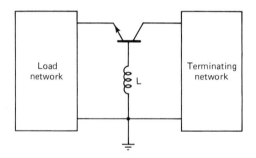

**Figure 5.5.2** Common-base configuration.

The two common network configurations for GaAs FETs oscillators are shown in Fig. 5.5.3. The common-gate configuration is used in low-power oscillator circuits since it is easy to tune. A series inductive feedback is usually required to improve $|\Gamma_{IN}|$ and $|\Gamma_{OUT}|$. The common-source configuration is used for higher oscillator output power and the feedback network is usually a capacitor. The common-drain configuration is not popular because the oscillator implementation is difficult.

A GaAs FET oscillator can also be built using the reverse-channel configuration shown in Fig. 5.5.4. A reverse-channel configuration uses a symmetrical GaAs FET with a negative voltage applied to the drain terminal. The transistor becomes a noninverting device, making the common lead inductance regenerative. The $S$ parameters in this reverse-channel configuration show that $|S_{12}|$ increases markedly with frequency and $|S_{11}|$ is greater than unity in a large frequency range.

Vendelin [5.6] shows some configurations, and discusses some design procedures, for low-noise oscillators and buffered oscillators.

The load tuning elements are not limited to lossless or $RLC$ networks.

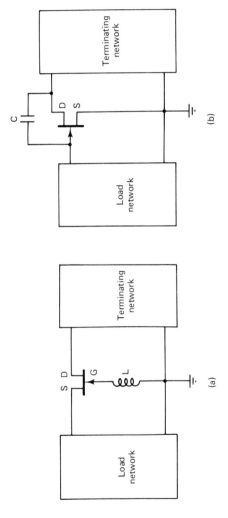

**Figure 5.5.3** (a) Common-gate configuration; (b) common-source configuration.

**Figure 5.5.4** Reverse-channel GaAs FET.

Load tuning networks can be designed using YIG (yttrium iron garnet) resonators, varactor diodes, and so on.

A YIG resonator consists of a ferrimagnetic material which can be modeled by a parallel *RLC* resonant circuit. The value of the elements depend on the magnetization, coupling, and resonance linewidth of the YIG sphere, and on the applied dc magnetic field. The uniform dc magnetic field is applied with an electromagnet with a single gap. The gap design is important since a nonuniform dc magnetic field results in a tuning hysteresis and spurious responses. A common-gate GaAs FET oscillator using a YIG resonator is shown in Fig. 5.5.5.

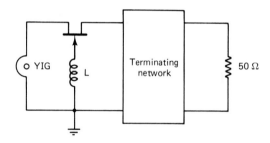

**Figure 5.5.5** YIG-tuned oscillator.

The YIG sphere is strongly coupled to the transmission line that connects to the active device. Assuming that the YIG sphere is always magnetically saturated and that the sphere diameter is $\ll \lambda/4$, the YIG device can be modeled by a parallel resonant circuit, as shown in Fig. 5.5.6. The element values are given by [5.7]

$$G_o = \frac{d^2}{\mu_o V \omega_m Q_U} + G_L$$

$$L_o = \frac{\mu_o V \omega_m}{\omega_o d^2}$$

$$C_o = \frac{1}{\omega_o^2 L_o}$$

where

$$\omega_m = \gamma 8\pi^2 M_s$$

$$Q_U = \frac{H_o - 4\pi M_s/3}{\Delta H}$$

Sec. 5.5  Oscillator Configurations

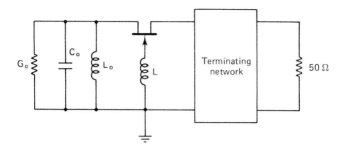

**Figure 5.5.6** Equivalent network of a YIG sphere in a YIG-tuned oscillator.

Here $4\pi M_s$ is the saturation magnetization of the sphere, $\mu_o = 4\pi(10^{-7})$ henrys per meter, $V$ is the volume of the sphere, $d$ the coupling loop diameter, $\gamma$ the gyromagnetic ratio (2.8 MHz/Oe), $H_o$ the applied dc magnetic field, $Q_U$ the unloaded $Q$, $\Delta H$ the resonance line width (approximately 0.2 Oe), and $\omega_o$ the center frequency of resonance. The frequency $\omega_o$ can be expressed as

$$\omega_o = 2\pi\gamma H_o$$

A varactor-tuned oscillator uses the voltage-controlled capacitance of a varactor diode to accomplish the electronic tuning. A basic schematic of a varactor-tuned oscillator is shown in Fig. 5.5.7.

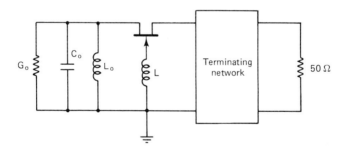

**Figure 5.5.7** Varactor-tuned oscillator.

Varactor diodes of different types having a wide range of capacitances are available. In the varactor circuit model shown in Fig. 5.5.8, the varactor diode capacitance ($C_v$), for Schottky-type devices, is given by the formula

$$C_v = \frac{C_0}{(1 + V/\phi)^{1/2}}$$

where $C_0$ is the value of capacitance at zero voltage, $V$ the reverse bias voltage, and $\phi$ the junction contact potential ($\phi \approx 0.7$ V). The resistance $R_s$ represents the series resistance of the diode, and the reverse diode resistance $R_r$ is large and therefore can be neglected.

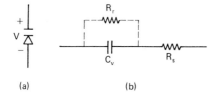

**Figure 5.5.8** (a) Varactor diode circuit symbol; (b) model.

## PROBLEMS

**5.1.** Design a 2-GHz oscillator using a BJT with external feedback as shown in Fig. 5.3.3. The S parameters of the network at 2 GHz are as follows:

|     | L = 0 H | L = 0.5 nH |
|-----|---------|------------|
| $S_{11}$ | 0.94 $\lfloor 174°$ | 1.04 $\lfloor 173°$ |
| $S_{21}$ | 1.90 $\lfloor -28°$ | 2.00 $\lfloor -30°$ |
| $S_{12}$ | 0.013 $\lfloor 98°$ | 0.043 $\lfloor 153°$ |
| $S_{22}$ | 1.01 $\lfloor -17°$ | 1.05 $\lfloor -18°$ |

(This problem is based on a design from Ref. [5.6].)

**5.2.** In Problem 5.1 implement the input tuning network using a YIG sphere. Specify the characteristics of the YIG sphere.

**5.3.** Design the input tuning network for the oscillator in Example 5.3.1.

**5.4.** Design a 10-GHz oscillator using a common-gate GaAs FET. The S parameters of the transistor at 10 GHz, $V_{DS} = 6$ V, $I_{DS} = 150$ mA, are

$$S_{11} = 0.85 \lfloor -36°$$
$$S_{21} = 0.53 \lfloor 96°$$
$$S_{12} = 0.22 \lfloor -36°$$
$$S_{22} = 1.125 \lfloor 171°$$

Show the dc bias network.

**5.5.** Implement the input tuning network of Example 5.3.2 using a varactor diode. Specify the diode characteristics.

**5.6.** Verify the relations (5.4.1) to (5.4.6).

**5.7.** Design an 8-GHz GaAs FET oscillator using the large-signal method discussed in Section 5.4. The S parameters of the transistor at 8 GHz are

$$S_{11} = 0.8 \lfloor 140°$$
$$S_{12} = 0.2 \lfloor -70°$$
$$S_{21} = 0.8 \lfloor 140°$$
$$S_{22} = 0.9 \lfloor 170°$$

The large-signal characteristics with the termination $Z_{T,o}$ are shown in Fig. P5.5.

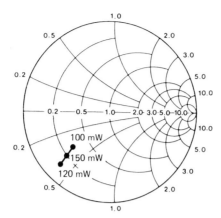

Figure P5.5

**5.8.** (a) Johnson [5.8] shows that the output power of an oscillator can be approximated with the equation

$$P_{OUT} = P_{sat}(1 - e^{-G_o P_{IN}/P_{sat}})$$

where $P_{sat}$ is the saturated output power of the amplifier, $P_{IN}$ is the input power, and $G_o$ is the small-signal power gain. Show that the maximum oscillator power [$P_{osc}(\max)$] is given by

$$P_{osc}(\max) = P_{sat}\left(1 - \frac{1}{G_o} - \frac{\ln G_o}{G_o}\right)$$

and the maximum efficient gain, defined by

$$G_{ME} = \frac{P_{OUT} - P_{sat}}{P_{IN}}$$

is given by

$$G_{ME}(\text{max. oscillator power}) = \frac{G_o - 1}{\ln G_o}$$

*Hint:* The maximum oscillator power occurs at the point of maximum $P_{OUT} - P_{IN}$, or where

$$\frac{\partial P_{OUT}}{\partial P_{IN}} = 1$$

(b) A GaAs FET has $G_o = 7.5$ dB with $P_{sat} = 1$ W. Calculate the maximum oscillator power and the corresponding maximum gain.

(c) Draw typical $P_{osc}/P_{sat}$ versus $G_o$, and $G_{ME}$ (maximum oscillator power) versus $G_o$ plots.

**5.9.** The conductance of a two-terminal negative-resistance device used in a YIG-tuned oscillator has a dependence on RF voltage amplitude $A$ given by $-G = (15 - 2A) \times 10^{-3}$ S, where $A < 7.5$ V. Assuming an equivalent circuit of the form shown in Fig. P5.9 plot the power variation of the oscillator when the dc magnetic field is increased from 3000 to 4500 Oe. Use $V = 0.75 \times 10^{-6} m^3$, $d = 2$ mm, $4\pi M_s = 1750$ G, $G_L = 10$ mS, $l = 10$ mm, $Z_o = 200$ Ω, and $\Delta H = 0.5$ Oe.

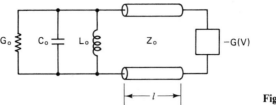

**Figure P5.9**

## REFERENCES

[5.1] K. Kurokawa, "Some Basic Characteristics of Broadband Negative Resistance Oscillator Circuits," *The Bell System Technical Journal*, July 1969.

[5.2] P. C. Wade, "Novel FET Power Oscillators," *Electronics Letters*, September 1978.

[5.3] P. C. Wade, "Say Hello to Power FET Oscillators," *Microwaves*, April 1979.

[5.4] COMPACT reference manual, Compact Software, Inc., 1131 San Antonio Road, Palo Alto, CA 94303.

[5.5] W. Wagner, "Oscillator Design by Device Line Measurement," *Microwave Journal*, February 1979.

[5.6] G. D. Vendelin, *Design of Amplifiers and Oscillators by the S Parameter Method*, Wiley-Interscience, New York, 1982.

[5.7] D. V. Morgan and M. J. Howes, editors, *Microwave Solid State Devices and Applications*, Peter Peregrinus Ltd., New York, 1980.

[5.8] K. M. Johnson, "Large Signal GaAs MESFET Oscillator Design," *IEEE Transactions on Microwave Theory and Techniques*, March 1979.

# APPENDIX A

## COMPUTER-AIDED DESIGN: COMPACT AND SUPER-COMPACT*

The use of computer analysis and optimization programs is of great importance in the design of microwave transistor amplifiers. In some designs, straightforward calculations are a tedious process, and a CAD program is a must. In fact, CAD methods are extremely valuable when used in conjunction with good engineering principles.

In this appendix, some of the capabilities of the programs COMPACT and SUPER-COMPACT [A.1] are described first, followed by an example. Although the program COMPACT is now considered to be an "old" program, the example is worked using both COMPACT and SUPER-COMPACT. These large-scale CAD programs are very powerful, providing a wide range of capabilities.

Among the features of COMPACT are: circuit analysis, stability analysis, sensitivity analysis, optimization (up to 15 variables), a Monte Carlo analysis that permits the designer to evaluate component tolerances, and so on. SUPER-COMPACT, introduced in 1980, provides all the features of COMPACT, plus a host of new capabilities. Some of the new capabilities of SUPER-COMPACT are an efficient circuit interconnection feature, total file flexibility, interactive graphics, better optimization, new circuit elements, and so on. The interactive graphics plots polar or rectangular charts of $S$ parameters, constant-noise circles, constant-gain and stability circles, and so on. The reader should refer to Ref. [A.1] for the full range of capabilities of these programs.

The following example illustrates the use of the large-scale CAD programs COMPACT and SUPER-COMPACT in the analysis, design, and optimization of a microwave transistor amplifier.

*COMPACT and SUPER-COMPACT are trademarks of Compact Software, Inc. [A.1].

## COMPACT

In Example 4.4.2 a preliminary analysis and design of a BJT broadband amplifier was performed. The network in Fig. 4.4.13 produced the $S$ parameters given in Fig. 4.4.14a. The results given in Fig. 4.4.14b were obtained by optimizing the values of $R_2$ and $L_2$ for $G_T = 10$ dB using COMPACT. The COMPACT program used to obtain the optimized values (i.e., $R_2 = 274.8 \ \Omega$ and $L_2 = 12.9$ nH) is

```
 1     TWO AA S1 50
 2     SRL JJ SE -208 -28
 3     PAR AA JJ
 4     RES BB PA 300
 5     CAS AA BB
 6     PRI AA S4 50
 7     END
 8     10 100 250 500 750 1000 1250 1500
 9     END
10     .95   -2 7.35 174.6 .003 84.3 1.01  -1
11     .92  -11 7.15 168.0 .007 79.0  .99  -4
12     .87  -28 6.83 154.5 .015 69.2  .96 -10
13     .78  -54 6.28 135.0 .026 54.0  .90 -18
14     .69  -78 5.67 123.0 .033 41.4  .84 -25
15     .63  -98 5.04 113.0 .037 33.0  .79 -30
16     .60 -114 4.42  99.9 .038 29.3  .77 -33
17     .60 -127 3.88  87.0 .039 28.0  .76 -35
18     END
19     .1
20     10 10 5 10
21     END
```

In COMPACT, the first column of a program statement is a three-letter abbreviation that describes the component. The second column is a two-letter name (both letters must be the same) assigned to the component. The third column supplies additional information about the component. A description of the program follows:

(a) Line 1 describes the transistor as a two-port network (TWO), called element AA. The $S$ parameters (given in lines 10 to 17) with the S1 option are given at the frequencies 10 to 1500 MHz (see line 8). The $S$-parameters reference impedance is indicated in the fourth column (i.e., 50 Ω).

(b) Line 2 describes the series resistance–inductance (SRL) element in the feedback path (see Fig. 4.4.13). In this element, called element JJ, $R_2 = 208 \ \Omega$ and $L_2 = 28$ nH. The minus sign calls for optimization of these elements.

(c) Line 3 directs the program to interconnect element AA [i.e., the transistor (TWO)] with element JJ (i.e., the SRL branch). Since the elements are connected in parallel, the parallel (PAR) interconnection is used. The resulting two-port element, after the PAR interconnection, is called AA.

(d) Line 4 shows that the 300-Ω resistor (RES), called element BB, is connected in parallel (PA).

(e) Line 5 directs the program to cascade (CAS) (i.e., to interconnect) the elements AA and BB. After cascading the elements, the resulting two-port is called element AA.

Appendix A: Computer-Aided Design: Compact and Super-Compact 219

(f) Line 6 directs the program to print (PRI) the S parameters of element AA, at the frequencies indicated in line 8. The S4 option also prints the reflection coefficients for a simultaneous conjugate match.

(g) Line 7 ends (END) the definition of the elements and the print out option selected.

(h) Line 8 gives the frequency information. The S parameters are given at 10, 100, 250, 500, 750, 1000, 1250, and 1500 MHz.

(i) Line 9 ends (END) the frequency information.

(j) Lines 10 to 17 provide the S-parameter information (i.e., $S_{11}$, $S_{21}$, $S_{12}$, and $S_{22}$) in polar form, at each frequency.

(k) Line 18 ends (END) the S-parameter data.

(l) Lines 19 and 20 provide the optimization data. Line 19 contains only one number, called TEF (i.e., the terminal error function). This value of TEF (i.e., 0.1) specifies the value of error function when the optimization is to be terminated.

(m) Line 20 defines the error function in the frequency band. The error function, called EF, in the band is of the form

$$\text{EF} = \frac{1}{N} \sum_{FI}^{FF} W_1 |S_{11}|^2 + W_2 |S_{22}|^2 + W_3 ||S_{21}(\text{dB})|^2 - G_T(\text{dB})|^2$$

where $N$ is the number of frequencies, FI and FF are the initial and final frequencies in the band, $W_1$, $W_2$, and $W_3$ are the weighting factors, and $G_T$(dB) is the desired gain in decibels.

The optimization process will try to minimize the value of EF. A large weighting factor will result in heavy emphasis on that part of the error function. Line 20 uses $W_1 = W_2 = 10$, $W_3 = 5$, and $G_T = 10$ dB. That is, there is a heavy emphasis in optimizing $S_{11}$, $S_{22}$, and $G_T$.

(n) Line 21 ends (END) the optimization data information.

The program was run and the output shown below resulted.

```
                CIRCUIT OPTIMIZATION WITH  2 VARIABLES
    INITIAL CIRCUIT ANALYSIS

                  POLAR S-PARAMETERS IN   50.0 OHM SYSTEM
      FREQ.     S11            S21            S12           S22        S21    K
              (MAGN<ANGL) ( MAGN<ANGL) ( MAGN<ANGL) (MAGN<ANGL)   DB    FACT.

       10.00   .08<   17   2.68< 176.7   .184<   1.0   .04<  140   8.58  1.25
      100.00   .07<   17   2.67< 175.5   .182<  -2.9   .07<  113   8.53  1.26
      250.00   .08<   11   2.73< 171.5   .180<  -9.6   .14<   93   8.74  1.23
      500.00   .10<   -5   3.03< 163.7   .172< -20.4   .26<   77   9.62  1.14
      750.00   .12<  -46   3.37< 156.6   .154< -32.6   .35<   62  10.56  1.11
     1000.00   .18<  -77   3.65< 146.3   .133< -43.6   .44<   45  11.25  1.10
     1250.00   .28<  -93   3.79< 128.9   .112< -54.4   .56<   26  11.58  1.08
     1500.00   .39< -109   3.70< 109.5   .087< -64.1   .65<    8  11.35  1.12
```

```
OPTIMIZATION BEGINS WITH FOLLOWING VARIABLES AND GRADIENTS
          VARIABLES                    GRADIENTS
   ( 1): 208.000               ( 1):-34.3745
   ( 2): 28.0000               ( 2): 14.5541
   ERR. F.=     9.388
           ----****----

   ERR. F.=     3.636
   ERR. F.=     1.410
   ERR. F.=     1.269
   ERR. F.=     1.241
   ( 1): 274.819               ( 1): .852160-002
   ( 2): 12.8690               ( 2): .357162-002
   ERR. F.=     1.240
           ----****----
GRADIENT TERMINATION WITH ABOVE VALUES. FINAL ANALYSIS FOLLOWS
   10.00   .18<    5   3.14< 176.3   .161<   1.0   .07<   20    9.95   1.22
  100.00   .18<   -4   3.11< 173.0   .159<  -1.9   .08<   32    9.86   1.23
  250.00   .17<  -22   3.10< 165.5   .156<  -6.8   .11<   45    9.83   1.25
  500.00   .18<  -52   3.17< 153.7   .148< -14.0   .17<   49   10.01   1.26
  750.00   .21<  -87   3.20< 145.8   .134< -21.2   .21<   44   10.11   1.33
 1000.00   .25< -110   3.19< 137.4   .120< -26.6   .25<   36   10.06   1.40
 1250.00   .30< -120   3.14< 124.2   .109< -31.2   .34<   25    9.95   1.45
 1500.00   .36< -128   3.06< 109.5   .098< -35.7   .43<   15    9.72   1.49
      POLAR COORDINATES OF SIMULTANEOUSCONJUGATE MATCH
      F       SOURCE REFL. COEFF.    LOAD REFL. COEFF.    GMAX
     MHZ        MAGN.<ANGLE            MAGN.<ANGLE         DB
    10.0         .21<  -11              .07<-131          10.12
   100.0         .20<   -5              .07<-116          10.01
   250.0         .18<    7              .10<  -96         10.00
   500.0         .16<   27              .15<  -76         10.23
   750.0         .15<   77              .15<  -53         10.38
  1000.0         .18<  113              .18<  -33         10.48
  1250.0         .20<  126              .27<  -21         10.64
  1500.0         .24<  133              .35<  -12         10.83
```

The initial circuit analysis for the amplifier resulted in poor performance in the passband (i.e., $G_T$ varies from 8.53 to 11.58 dB). Observe that the column $S_{21}$ in decibels represents the quantity $10 \log |S_{21}|^2$, or $G_T$ in a 50-$\Omega$ system. The error function of the original amplifier is observed to be 9.388. The optimization procedure reduced the error function to 1.24. The optimized element values are listed above this final error function (i.e., $R_2 = 274.8 \, \Omega$ and $L_2 = 12.8$ nH).

COMPACT has a built-in test for the process of optimization. If the error function is not reduced by a certain amount during consecutive iterations, the search is terminated (FRACTIONAL TERMINATION) followed by a final analysis. Also, if the gradients reach such low values that the numerical capabilities of the computer result in errors, the search is terminated (GRADIENT TERMINATION) followed by a final analysis.

A two-section matching network at the input and output can now be designed to provide gain flatness and good VSWR. From the previous data at 1500 MHz, $\Gamma_{Ms} = 0.24 \lfloor 133°$ and $\Gamma_{ML} = 0.35 \lfloor -12°$, and the matching networks shown in Fig. A.1 are obtained. The amplifier shown in Fig. A.1 can then be optimized, using the following COMPACT program:

```
     1     CAP AA PA -1.46
     2     IND BB SE -3.77
     3     TWO CC S1 50
     4     SRL JJ SE -274.8 -12.9
     5     PAR CC JJ
     6     RES DD PA 300
     7     CAP EE PA -1.23
     8     IND FF SE -5.41
     9     CAX AA FF
    10     PRI AA S1 50
    11     END
```

# Appendix A: Computer-Aided Design: Compact and Super-Compact

```
12        10 100 250 500 750 1000 1250 1500
13        END
14        .95   -2  7.35 174.6 .003 84.3 1.01  -1
15        .92  -11  7.15 168.0 .007 79.0  .99  -4
16        .87  -28  6.83 154.5 .015 69.2  .96 -10
17        .78  -54  6.28 135.0 .026 54.0  .90 -18
18        .69  -78  5.67 123.0 .033 41.4  .84 -25
19        .63  -98  5.04 113.0 .037 33.0  .79 -30
20        .60 -114  4.42  99.9 .038 29.3  .77 -33
21        .60 -127  3.88  87.0 .039 28.0  .76 -35
22        END
23        .1
24        10 10 5 10
25        END
```

The lines that need description in this program are:

(a) Line 1 shows that the first element is a capacitor (CAP), called AA, connected in parallel. The value of the capacitor is 1.46 pF. Line 7 describes the 1.23-pF capacitor, and is similar to line 1.

(b) Line 2 shows that the second element is an inductor (IND), called element BB, connected in series. The value of the inductor is 3.77 nH. Line 8 describes the 5.41-nH inductor.

(c) Line 9 is a code (CAX) that directs the program to cascade the elements in the given order (i.e., from AA to FF).

(d) The minus signs in lines 1, 2, 4, 7, and 8 request optimization of six variables.

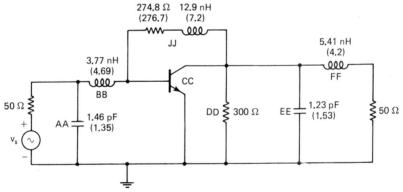

**Figure A.1** Schematic of a broadband amplifier. The optimized values are shown in parentheses.

The program was run and the output shown below resulted.

```
                  CIRCUIT OPTIMIZATION WITH  6 VARIABLES
   INITIAL CIRCUIT ANALYSIS

                  POLAR S-PARAMETERS IN   50.0 OHM SYSTEM
       FREQ.        S11           S21           S12           S22        S21     K
                (MAGN<ANGL) ( MAGN<ANGL) ( MAGN<ANGL) (MAGN<ANGL)   DB   FACT.

       10.00    .18<    5   3.14< 175.7   .161<    .5   .07<   21   9.95  1.22
      100.00    .18<  -12   3.11< 167.3   .159<  -7.6   .09<   35   9.85  1.23
      250.00    .17<  -40   3.09< 151.4   .155< -20.9   .13<   43   9.80  1.25
      500.00    .17<  -81   3.15< 125.9   .147< -41.8   .19<   39   9.96  1.26
      750.00    .16< -126   3.21< 104.1   .134< -63.0   .20<   31  10.14  1.32
     1000.00    .13< -163   3.28<  81.1   .123< -82.9   .17<   24  10.31  1.40
     1250.00    .07<  171   3.38<  52.0   .117<-103.4   .11<    7  10.57  1.45
     1500.00    .01< -167   3.48<  18.7   .111<-126.5   .01< -101  10.83  1.49
```

```
OPTIMIZATION BEGINS WITH FOLLOWING VARIABLES AND GRADIENTS
        VARIABLES                 GRADIENTS
     ( 1): 1.46000             ( 1): .977702-001
     ( 2): 3.77000             ( 2):-.424543-001
     ( 3): 274.800             ( 3): 1.87408
     ( 4): 12.9000             ( 4): 3.36864
     ( 5): 1.23000             ( 5): .169268-001
     ( 6): 5.41000             ( 6): .503194-001
     ERR. F.=       1.147
               ----****----


     ( 1): 1.35040             ( 1): .346685-002
     ( 2): 4.68527             ( 2):-.305893-001
     ( 3): 276.676             ( 3):-.341341-001
     ( 4): 7.19841             ( 4): .994187-003
     ( 5): 1.52804             ( 5):-.230386-002
     ( 6): 4.19881             ( 6):-.112352-001
     ERR. F.=        .301
               ----****----
FRACTIONAL TERMINATION WITH ABOVE VALUES. FINAL ANALYSIS FOLLOWS
   10.00  .19<    5  3.16<  175.7  .160<    .5   .07<    18   9.98  1.22
  100.00  .18<   -9  3.12<  166.9  .158<  -7.2   .08<    22   9.88  1.23
  250.00  .17<  -34  3.10<  150.3  .155< -20.1   .09<    27   9.83  1.25
  500.00  .15<  -74  3.15<  123.4  .148< -40.5   .11<    25   9.97  1.28
  750.00  .12< -128  3.19<  100.0  .136< -61.5   .08<    14  10.08  1.35
 1000.00  .09<  170  3.19<   75.3  .126< -81.7   .03<     1  10.07  1.44
 1250.00  .08<  101  3.18<   44.9  .118<-102.4   .03<  -127  10.05  1.50
 1500.00  .12<   51  3.14<   10.9  .111<-125.3   .11<  -164   9.94  1.56
```

The final analysis shows that the transducer power gain, over the band, varies from 9.83 to 10.08 dB (i.e., an excellent gain flatness). The input and output VSWR are seen to be better than 1.5. The optimized values of the elements are shown in parentheses in Fig. A.1.

## SUPER-COMPACT

The optimization of the amplifier in Example 4.4.2 is now performed using SUPER-COMPACT. In order to show a more impressive optimization, some of the component values used in the original amplifier are selected somewhat arbitrarily. These values are shown in Fig. A.2. Figure A.2 also shows the node assignment used in the program.

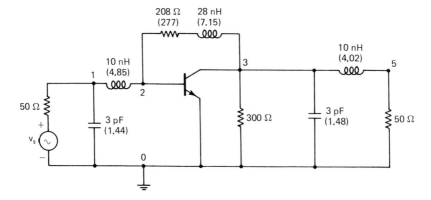

**Figure A.2** Schematic of the initial and optimized (values in parentheses) broadband amplifier.

Appendix A: Computer-Aided Design: Compact and Super-Compact

The following SUPER-COMPACT program was used to analyze and optimize the amplifier in Fig. A.2.

```
BLK
   TWO 2 3 0 Q1
   SRL 2 3 R=?208? L=?28NH?
   FB: 2POR 2 3
END
LAD
   CAP 1 0 C=?3PF?
   IND 1 2 L=?10NH?
   FB 2 3
   RES 3 0 R=300
   CAP 3 0 C=?3PF?
   IND 3 5 L=?10NH?
   AMP: 2POR 1 5
END
FREQ
   1E7 1E8 STEP 250MHZ 1.5GHZ 250MHZ
END
OUT
   PRI AMP S
END
OPT
   AMP MS11=0 W=10  MS22=0 W=10  MS21=10DB W=5
   TERM=0.1
END
DATA
   Q1: S
   1E7    .95  -2    7.35 175   .003 84   1.01  -1
   1E8    .92  -11   7.15 168   .007 79    .99  -4
   5E8    .78  -54   6.28 135   .026 54    .90 -10
   1E9    .63  -98   5.04 113   .037 33    .79 -30
   1.5E9  .60 -127   3.88  87   .039 28    .76 -35
END
```

A brief description of the program follows:

(a) In SUPER-COMPACT the input file is divided into blocks. Each block ends with "END". The first block (i.e., from BLK to END) describes the transistor as a two-port network (called TWO), and the series resistance–inductance combination (SRL) element in the feedback path. The element TWO is connected between nodes 2, 3, and 0. The label Q1 establishes a reference to the DATA block, where the $S$ parameters for TWO are given. The ground node is always zero. The element SRL is connected between nodes 2 and 3. In this element the initial values are $R = 208$ Ω and $L = 28$ nH. The question marks call for optimization of these elements. FB is the label selected for the two-port (2POR) network formed by TWO and SRL. The two-port FB appears between nodes 2 and 3 (with 0 as ground).

(b) The second block (i.e., from LAD to END) describes the 3-pF capacitor (CAP) connected between nodes 1 and 0, the 10-nH inductor (IND) between nodes 1 and 2, the FB network [i.e., the resulting two port from (a)] between nodes 2 and 3, the 300-Ω resistor (RES) between nodes 3 and 0, the 3-pF capacitor between nodes 3 and 0, and the 10-nH inductor between nodes 3 and 5. AMP is the label selected for the resulting two-port (2POR) network between nodes 1 and 5 (with 0 as ground). The elements with question marks are to be optimized.

(c) The third block (i.e., from FREQ to END) gives the frequency information. The frequencies given are 10 MHz (1E7), 100 MHz, and from 250 MHz to 1.5 GHz in 250-MHz steps.

(d) The fourth block (i.e., from OUT to END) directs the program to print (PRI) the two-port $S$ parameters (S) of AMP.

(e) The fifth block (i.e., from OPT to END) provides the optimization data. AMP is the name of the network to be optimized. MS11, MS22, and MS21 are the keywords for the magnitudes of $S_{11}$, $S_{22}$, and $|S_{21}|^2$. The optimization goal for $|S_{11}|$ is zero (i.e., MS11 = 0) with a weighting factor of 10, for $|S_{22}|$ the goal is zero with a weighting factor of 10, and

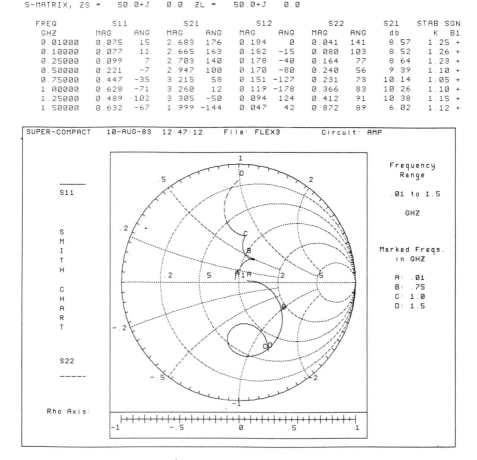

**Figure A.3** Initial analysis of the broadband amplifier.

Appendix A: Computer-Aided Design: Compact and Super-Compact 225

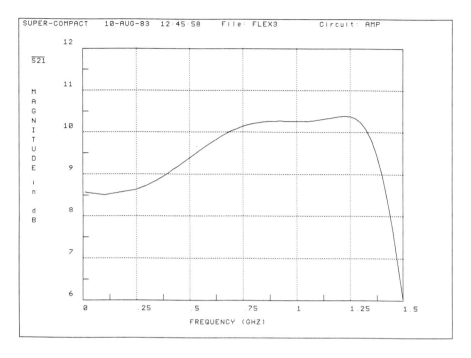

**Figure A.3** (*continued*)

for $|S_{21}|^2$ the goal is 10 dB with a weighting factor of 5. The value of TERM = 0.1 specifies the value of the error function when the optimization is to be terminated.

(f) The sixth block (i.e., from DATA to END) provides the S-parameter data for the transistor (TWO). The DATA block is partitioned by means of previously defined data labels. In this program the only label is Q1. The S parameters for the transistor are listed after Q1. Since the frequency block specifies eight frequencies and the S-parameter data are provided at five frequencies, the program uses linear interpolation to calculate the other S parameters.

The program was run and the analysis of the initial amplifier is shown in Fig. A.3. ZS and ZL are the values of the source and load impedances (i.e., 50 Ω). The plotting capabilities of SUPER-COMPACT were used to obtain the plots of $S_{11}$, $S_{22}$, and $|S_{21}|^2$ in decibels shown in Fig. A.3. Observe that the program calculates both $K$ and the sign of $B_1$ to determine the stability of the two-port.

As seen from Fig. A.3, the analysis of the original amplifier results in poor performance in the passband (i.e., $G_T$ varies from 6.02 to 10.38 dB), and the input and output VSWR are very poor.

A random-search optimization is now performed. A random-search optimization works well when the starting values of the variables are far from the optimum values. The random-search optimization produces the following results:

```
<OPTIM> CPU TIME =    2.86 SECS.
BEST VALUES:
VAR( 1) =    275.32
VAR( 2) =    0.13443E-07
VAR( 3) =    0.17460E-11
VAR( 4) =    0.35452E-08
VAR( 5) =    0.15248E-11
VAR( 6) =    0.80739E-08
V  =   1.0535E+00     21 FUNCTION EVALUATIONS     21 CALLS TO <COSTT>

CIRCUIT:  AMP
S-MATRIX, ZS =   50.0+J    0.0   ZL =   50.0+J    0.0

FREQ         S11              S21              S12              S22           S21    STAB  SGN
GHZ       MAG    ANG      MAG    ANG      MAG    ANG      MAG    ANG      db       K    B1
0.01000   0.185    3     3.147   176    0.160     0     0.069    20     9.96    1.22  +
0.10000   0.178  -16     3.108   166    0.158    -9     0.095    40     9.85    1.23  +
0.25000   0.176  -47     3.074   148    0.154   -23     0.175    50     9.75    1.24  +
0.50000   0.181  -91     3.099   119    0.146   -45     0.291    43     9.83    1.23  +
0.75000   0.163 -128     3.177    94    0.135   -70     0.272    34    10.04    1.28  +
1.00000   0.104 -157     3.267    68    0.122   -97     0.180    47    10.28    1.40  +
1.25000   0.049  -96     3.314    34    0.110  -124     0.201    89    10.41    1.49  +
1.50000   0.203  -97     3.290    -8    0.103  -154     0.339   103    10.35    1.48  +
```

The optimized values of the elements are listed under BEST VALUES. For example, VAR(1) = 275.32 is the optimized value of the initial 208-$\Omega$ resistor.

Next, a gradient optimization can be performed. A gradient search is excellent when there is no local minimum and if the search is started reason-

```
OPTIMIZATION BEGINS WITH FOLLOWING VARIABLES AND GRADIENTS

          VARIABLES                GRADIENTS
       ( 1):   275.32           ( 1):-0.84043
       ( 2):   0.13443E-07      ( 2):  2.0659
       ( 3):   0.17460E-11      ( 3):  0.15259
       ( 4):   0.35452E-08      ( 4):-0.10610
       ( 5):   0.15248E-11      ( 5):-0.69737
       ( 6):   0.80739E-08      ( 6):-0.25630
       ERR. F. =        1.053
         ----****----
NUMBER OF ITERATIONS? (X/<0>): 10,0
       ERR. F. =        0.947
       ERR. F. =        0.865
       ERR. F. =        0.426
       ERR. F. =        0.381
       ERR. F. =        0.370
       ERR. F. =        0.353
       ERR. F. =        0.350
       ( 1):   277.23           ( 1):-0.74804E-01
       ( 2):   0.71540E-08      ( 2):-0.23544E-01
       ( 3):   0.14400E-11      ( 3):  0.59605E-02
       ( 4):   0.48502E-08      ( 4):  0.37551E-01
       ( 5):   0.14809E-11      ( 5):-0.30398E-01
       ( 6):   0.40249E-08      ( 6):  0.22948E-01
       ERR. F. =        0.346
         ----****----
FRACTIONAL TERMINATION WITH ABOVE VALUES.  FINAL ANALYSIS FOLLOWS
```

**Figure A.4** Gradient optimization and final analysis of the broadband amplifier.

## Appendix A: Computer-Aided Design: Compact and Super-Compact

```
CIRCUIT:   AMP
S-MATRIX, ZS =   50.0+J    0.0   ZL =   50.0+J    0.0
```

| FREQ | S11 | | S21 | | S12 | | S22 | | S21 | STAB | SGN |
|---|---|---|---|---|---|---|---|---|---|---|---|
| GHZ | MAG | ANG | MAG | ANG | MAG | ANG | MAG | ANG | db | K | B1 |
| 0.01000 | 0.188 | 4 | 3.159 | 176 | 0.160 | 0 | 0.070 | 16 | 9.99 | 1.22 | + |
| 0.10000 | 0.180 | -10 | 3.122 | 167 | 0.158 | -7 | 0.076 | 22 | 9.89 | 1.23 | + |
| 0.25000 | 0.170 | -33 | 3.101 | 150 | 0.155 | -19 | 0.107 | 36 | 9.83 | 1.25 | + |
| 0.50000 | 0.151 | -72 | 3.153 | 122 | 0.150 | -38 | 0.165 | 36 | 9.97 | 1.25 | + |
| 0.75000 | 0.118 | -119 | 3.195 | 99 | 0.140 | -59 | 0.129 | 20 | 10.09 | 1.32 | + |
| 1.00000 | 0.087 | 172 | 3.193 | 75 | 0.125 | -82 | 0.037 | -9 | 10.08 | 1.44 | + |
| 1.25000 | 0.106 | 89 | 3.158 | 46 | 0.114 | -104 | 0.049 | -160 | 9.99 | 1.54 | + |
| 1.50000 | 0.127 | 38 | 3.141 | 10 | 0.111 | -126 | 0.095 | -155 | 9.94 | 1.56 | + |

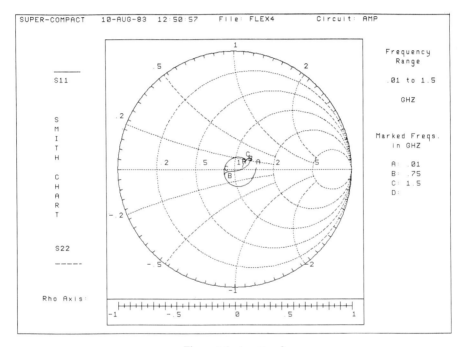

**Figure A.4** (*continued*)

ably close to the optimum. The results from the gradient optimization, together with the plots of $S_{11}$, $S_{22}$, and $|S_{21}|^2$ in decibels, are shown in Fig. A.4. An expanded view of the $S_{11}$ and $S_{22}$ plots is given in Fig. A.4.

The error function from the random-search optimization is 1.053. After the gradient optimization the error function was decreased to 0.346. The final analysis shows that the transducer power gain over the band varies from 9.83 to 10.09 dB (i.e., an excellent gain flatness). The input and output VSWR are seen to be better than 1.5. The optimized values of the elements are shown in parentheses in Fig. A.2.

If noise figure data for the transistor are available, a noise figure analysis and optimization of the feedback amplifier can also be performed using SUPER-COMPACT.

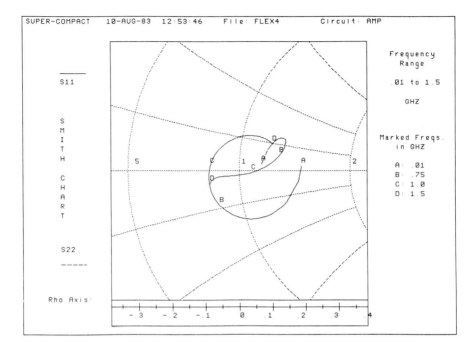

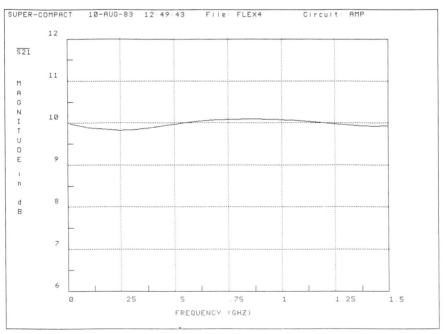

**Figure A.4** (*continued*)

Appendix A:  Computer-Aided Design: Compact and Super-Compact

## REFERENCES

[A.1] COMPACT and SUPER-COMPACT (computer programs), Compact Software, Inc., 1131 San Antonio Road, Palo Alto, CA 94303.

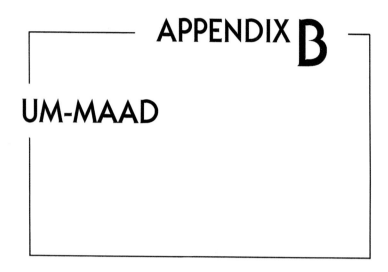

# APPENDIX B

# UM-MAAD

This appendix describes the program UM-MAAD (University of Miami–Microwave Amplifier Analysis and Design). UM-MAAD was developed to provide a simple CAD program that can be used in the analysis and design of some microwave transistor amplifiers. Also, UM-MAAD was structured so that it can be modified and easily expanded by the user. For example, the user can add subroutines to perform cascade connections of two-ports, to design oscillator networks, to treat more than one set of $S$ parameters, to calculate the dimensions and characteristics of microstrip lines, and so on.

The program was written in Fortran-77 and should run on most Fortran compilers with little or no modifications. The program interacts with the user; therefore, once it is running, the user should simply follow the instructions appearing on the computer terminal.

The program consists of a main program, called UM-MAAD, ten subroutines, and three functions. A brief description of the program follows.

### Program UM-MAAD

This is the main program. First, the program calls subroutine ENTER to read the $S$ parameters of a transistor and calculates some quantities that are common to other subroutines. Then the program calls subroutine SELECT, which selects the type of calculations to be performed. The type of calculations are determined by the integer $N$ from SELECT.

Appendix B  UM-MAAD

### Subroutine ENTER

ENTER reads the $S$ parameters in polar form and calculates $K$, $\Delta$, $C_1$, and $C_2$ from (3.3.13), (3.3.17), (3.8.3), and (3.8.10).

### Subroutine SELECT

SELECT prints and requests the type of calculation to be performed. SELECT reads the number $N$, where $N = 1$ to 9, that directs UM-MAAD to perform a specific calculation.

### Subroutine STABIL

STABIL calculates the input and output stability circles from (3.3.7) to (3.3.10), and determines the stable and unstable regions in the Smith chart.

### Subroutine MERIT

MERIT calculates the unilateral figure of merit and determines the range of $G_T/G_{TU}$ from (3.5.2) and (3.5.3).

### Subroutine UNGAIN

UNGAIN calculates $G_{s,\max}$, $G_{L,\max}$, $G_o$, and $G_{TU,\max}$ using (3.4.2) and (3.2.12). It also calculates the unilateral constant-gain circles for unconditionally stable and potentially unstable transistors. The calculations are based on (3.4.3) and (3.4.7) to (3.4.9).

### Subroutine SIMAT

SIMAT calculates the reflection coefficients $\Gamma_{Ms}$ and $\Gamma_{ML}$ (i.e., the simultaneous conjugate match) for an unconditionally stable transistor using (3.6.5) to (3.6.8).

### Subroutine POWERG

POWERG calculates constant operating power gain and available power gain circles. The operating power gain circles are calculated using (3.8.1), (3.8.5), and (3.8.6), and the available power gain circles from (3.8.9) to (3.8.12). The maximum stable gain is calculated using (3.6.11).

### Subroutine NOISEF

NOISEF calculates constant noise figure circles using (4.3.5), (4.3.7), and (4.3.8).

### Subroutine REFLEC

REFLEC calculates input and output reflection coefficients using (3.2.5) and (3.2.6).

### Subroutine CONVER

CONVER calculates the conversion from a normalized impedance to a reflection coefficient, or from a reflection coefficient to a normalized impedance or admittance, using (2.2.2).

### Function CANG

CANG calculates the angle in radians of a complex number.

### Function ANG

ANG converts an angle in radians to an angle in degrees.

### Function RAD

RAD converts an angle in degrees to an angle in radians.

## PROGRAM LISTING

This program was developed by Ching Y. Kung and Guillermo Gonzalez.

```
 1    C        UM-MAAD
 2             COMPLEX S11,S21,S12,S22,DELTA,C1,C2
 3             CALL ENTER (S11,S21,S12,S22,DELTA,DM,DA,AK,C1,C2)
 4    1        CALL SELECT (N)
 5             GO TO (10,20,30,40,50,60,70,80,90),N
 6    10       CALL STABIL (S11,S21,S12,S22,DM,DA,AK,C1,C2)
 7             GO TO 1
 8    20       CALL MERIT (S11,S21,S12,S22)
 9             GO TO 1
10    30       CALL UNGAIN (S11,S21,S22)
11             GO TO 1
12    40       CALL SIMAT (S11,S22,DM,AK,C1,C2)
13             GO TO 1
14    50       CALL POWERG (S11,S21,S12,S22,DM,AK,C1,C2)
15             GO TO 1
16    60       CALL NOISEF
17             GO TO 1
18    70       CALL REFLEC (S11,S21,S12,S22)
19             GO TO 1
20    80       CALL CONVER
21             GO TO 1
22    90       STOP
23             END
```

Appendix B    UM-MAAD                                                                    **233**

```
1              SUBROUTINE ENTER (S11,S21,S12,S22,DELTA,DM,DA,AK,C1,C2)
2              COMPLEX S11,S21,S12,S22,DELTA,C1,C2
3       9      PRINT *,' ENTER MAG. AND ANG. OF S11,S21,S12,S22'
4              READ *,S11M,S11A,S21M,S21A,S12M,S12A,S22M,S22A
5              S11=CMPLX(S11M*COS(RAD(S11A)),S11M*SIN(RAD(S11A)))
6              S21=CMPLX(S21M*COS(RAD(S21A)),S21M*SIN(RAD(S21A)))
7              S12=CMPLX(S12M*COS(RAD(S12A)),S12M*SIN(RAD(S12A)))
8              S22=CMPLX(S22M*COS(RAD(S22A)),S22M*SIN(RAD(S22A)))
9              WRITE(6,19)
10      19     FORMAT(6X,'S11=',16X,'S21=',16X,'S12=',16X,'S22=')
11             WRITE(6,29)S11M,S11A,S21M,S21A,S12M,S12A,S22M,S22A
12      29     FORMAT(8(2X,F8.3))
13             DELTA=S11*S22-S12*S21
14             DM=CABS(DELTA)
15             DA=CANG(DELTA)
16             DA=ANG(DA)
17             C1=S11-DELTA*CONJG(S22)
18             C2=S22-DELTA*CONJG(S11)
19             H1=CABS(S21*S12)
20             IF (H1.EQ.0.) GO TO 39
21             AK=(1-CABS(S11)**2-CABS(S22)**2+DM**2)/(2*H1)
22             GO TO 49
23      39     AK=1.E20
24      49     RETURN
25             END


1              SUBROUTINE SELECT (N)
2              WRITE(6,11)
3       11     FORMAT(//'      ENTER 1 FOR STABILITY CIRCLES,K, AND DELTA'/
4              1'              2 FOR UNILATERAL FIGURE OF MERIT'/
5              2'              3 FOR UNILATERAL POWER GAIN CIRCLES'/
6              3'              4 FOR SIMULTANEOUS CONJUGATE MATCH'/
7              4'              5 FOR POWER & AVAIL. POWER GAIN CIRCLES'/
8              5'              6 FOR NOISE FIGURE CIRCLES'/
9              6'              7 FOR INPUT/OUTPUT REFLECTION COEFFICIENTS'/
10             7'              8 FOR CONVERSION BETWEEN GAMMA AND Z'/
11             8'              9 TO EXIT      ')
12             READ *,N
13             RETURN
14             END
```

```
1          SUBROUTINE STABIL (S11,S21,S12,S22,DM,DA,AK,C1,C2)
2          COMPLEX S11,S12,S21,S22,CL,CS,C1,C2
3          WRITE(6,21)DM,DA,AK
4       21 FORMAT(//,5X,'DELTA.MAG= ',F7.3,5X,'DELTA.ANG= ',
5          1F8.3,5X,'    K= ',F7.3,//)
6          RLD=CABS(S22)**2-DM**2
7          RL=CABS(S21*S12/RLD)
8          CL=CONJG(C2)/RLD
9          CLM=CABS(CL)
10         CLA=ANG(CANG(CL))
11         RSD=CABS(S11)**2-DM**2
12         RS=CABS(S12*S21/RSD)
13         CS=CONJG(C1)/RSD
14         CSM=CABS(CS)
15         CSA=ANG(CANG(CS))
16         IF((DM.LT.1.).AND.(AK.GE.1.)) GO TO 101
17         PRINT *,' POTENTIALLY UNSTABLE TRANSISTOR '
18         PRINT *,' --------------------------------'
19         GO TO 111
20     101 PRINT *,' UNCONDITIONALLY STABLE TRANSISTOR '
21         PRINT *,' ----------------------------------'
22     111 PRINT *
23         PRINT *,'**** INPUT STABILITY CIRCLE ****'
24         PRINT *
25         PRINT *,'RADIUS=',RS,'CENTER. MAG=',CSM,'CENTER.ANG=',CSA
26         IF(CABS(S22)-1.) 41,51,51
27      41 IF (CSM.GT.RS) GO TO 121
28         PRINT *,'INSIDE IS STABLE'
29         GO TO 131
30     121 PRINT *,'OUTSIDE IS STABLE'
31     131 GO TO 61
32      51 IF(CSM.GT.RS) GO TO 141
33         PRINT *,' OUTSIDE IS STABLE'
34         GO TO 61
35     141 PRINT *,'INSIDE IS STABLE'
36      61 PRINT *
37         PRINT *,'**** OUTPUT STABILITY CIRCLE ****'
38         PRINT *
39         PRINT *,' RADIUS=',RL,'CENTER.MAG=',CLM,'CENTER.ANG=',CLA
40         IF(CABS(S11)-1.) 81,91,91
41      81 IF(CLM.GT.RL)GO TO 211
42         PRINT *,' INSIDE IS STABLE'
43         GO TO 221
44     211 PRINT *,' OUTSIDE IS STABLE'
45     221 GO TO 301
46      91 IF(CLM.GT.RL) GO TO 231
47         PRINT *,' OUTSIDE IS STABLE'
48         GO TO 301
49     231 PRINT *,' INSIDE IS STABLE'
50     301 RETURN
51         END



1          SUBROUTINE MERIT (S11,S21,S12,S22)
2          COMPLEX S11,S21,S12,S22
3          UN=CABS(S12)*CABS(S21)*CABS(S11)*CABS(S22)
4          UD=(1-CABS(S11)**2)*(1-CABS(S22)**2)
5          U=UN/UD
6          UL=10*ALOG10((1+U)**(-2))
7          UR=10*ALOG10((1-U)**(-2))
8          PRINT *
9          PRINT*,' ****** UNILATERAL FIGURE OF MERIT = ',U
10         PRINT *
11         PRINT *, UL,' DB < GT/GTU < ',UR,' DB'
12         RETURN
13         END
```

Appendix B    UM-MAAD                                                    235

```
1           SUBROUTINE UNGAIN (S11,S21,S22)
2           COMPLEX S11,S21,S22,S
3           IF ((CABS(S11).GE.1.) .OR. (CABS(S22).GE.1.)) GO TO 82
4           GSM=10*ALOG10(1/(1-CABS(S11)**2))
5           GLM=10*ALOG10(1/(1-CABS(S22)**2))
6           GO=10*ALOG10(CABS(S21)**2)
7           GUMAX=GSM+GLM+GO
8           PRINT *
9           PRINT *,'    GSMAX(DB)= ',GSM,'    GLMAX(DB)= ',GLM
10          PRINT *,'    GO   (DB)= ',GO,'    GTUMAX(DB)= ',GUMAX
11          PRINT *
12          GO TO 2
13      82  PRINT *
14          PRINT *,'     POTENTIALLY UNSTABLE TRANSISTOR  '
15          PRINT *
16      2   PRINT *,'ENTER 1 FOR GS CIRCLE, 2 FOR GL CIRCLE, 3 TO EXIT'
17          READ *,M
18          IF (M.EQ.3) GO TO 92
19          PRINT *,' ENTER UNILATERAL GAIN IN DBS--START,STOP,STEP'
20          READ *,START,STOP,STEP
21          GO TO (12,22) M
22          PRINT *,' TRY AGAIN-ERROR '
23          GO TO 2
24      12  S=S11
25          PRINT *
26          PRINT *,'    ****** GS CIRCLES ******'
27          GO TO 32
28      22  S=S22
29          PRINT *
30          PRINT *,'    ****** GL CIRCLES ******'
31      32  AINC=ANG(CANG(1/S))
32          G=START
33      42  GI=10**(G/10)
34          GI=GI*(1-CABS(S)**2)
35          RISD=1-CABS(S)**2*(1-GI)
36          RISN=SQRT(1-GI)*(1-CABS(S)**2)
37          RIS=RISN/RISD
38          DIS=GI*CABS(S)/(1-CABS(S)**2*(1-GI))
39          PRINT *
40          PRINT *, G,' DB GAIN CIRCLE ', ' RADIUS = ',RIS
41          PRINT *,'     CENTER.MAG = ',DIS,'   CENTER.ANG = ',AINC
42          PRINT *
43          G=G+STEP
44          IF (G.LE.STOP) GO TO 42
45          GO TO 2
46      92  RETURN
47          END


1           SUBROUTINE SIMAT (S11,S22,DM,AK,C1,C2)
2           COMPLEX S11,S22,C1,C2,GAMMAS,GAMMAL
3           IF ((DM.LT.1.).AND.(AK.GT.1.)) GO TO 37
4           PRINT *,'POT. UNSTABLE TRANSISTOR, USE OPERATING'
5           PRINT *,'OR AVAILABLE POWER GAIN DESIGN'
6           GO TO 47
7       37  B1=1+CABS(S11)**2-CABS(S22)**2-DM**2
8           B2=1+CABS(S22)**2-CABS(S11)**2-DM**2
9           GAMMAS=(B1-SQRT(B1**2-4*CABS(C1)**2))/(2*C1)
10          GAMMAL=(B2-SQRT(B2**2-4*CABS(C2)**2))/(2*C2)
11          GSMAG=CABS(GAMMAS)
12          GSANG=ANG(CANG(GAMMAS))
13          GLMAG=CABS(GAMMAL)
14          GLANG=ANG(CANG(GAMMAL))
15          PRINT *
16          PRINT *,'  **** COMPLEX CONJUGATE MATCH AT THE INPUT PORT'
17          PRINT *,'      GAMMAS.MAG = ',GSMAG,'  GAMMAS.ANG = ',GSANG
18          PRINT *
19          PRINT *,'  **** COMPLEX CONJUGATE MATCH AT THE OUTPUT PORT '
20          PRINT *,'      GAMMAL.MAG = ',GLMAG,'  GAMMAL.ANG = ',GLANG
21      47  RETURN
22          END
```

```
1           SUBROUTINE POWERG (S11,S21,S12,S22,DM,AK,C1,C2)
2           COMPLEX S,S11,S21,S12,S22,C1,C2,C,CENTER
3           PRINT *
4           IF((AK.GT.1.).AND.(DM.LT.1.)) GO TO 93
5           GMSG=CABS(S21/S12)
6           GMSG=10*ALOG10(GMSG)
7           PRINT *,'     POTENTIALLY UNST. TRANSISTOR, MSG=',GMSG ,' DB'
8           GOTO 3
9     93    GPMAX=(AK-SQRT(AK**2-1))*CABS(S21/S12)
10          GPMAX=10*ALOG10(GPMAX)
11          PRINT *,'  UNCOND. STABLE TRAN. GPMAX=',GPMAX,' DB'
12    3     PRINT *
13          PRINT *,'    ENTER 1 FOR OPERATING GAIN CIRCLES. '
14          PRINT *,'          2 FOR AVAILABLE GAIN CIRCLES. '
15          PRINT *,'          3 FOR EXIT '
16          READ *,M
17          IF (M.EQ.3) GOTO 103
18          PRINT *,' ENTER GAIN IN DBS -- START,STOP,STEP'
19          READ *,START,STOP,STEP
20          GO TO  (13,23) M
21          PRINT *,' TRY AGAIN'
22          GOTO 3
23    13    S=S22
24          C=C2
25          PRINT *,'  ****** OPERATING POWER GAIN CIRCLES ******'
26          GO TO 33
27    23    S=S11
28          C=C1
29          PRINT *,'  ****** AVAILABLE POWER GAIN CIRCLES ******'
30    33    TMP1=CABS(S12*S21)
31          TMP2=CABS(S)**2-DM**2
32          G=START
33    43    GI=10**(G/10)/CABS(S21)**2
34          CENTER=GI*CONJG(C)/(1+GI*TMP2)
35          DISTAN=CABS(CENTER)
36          ANGLE=ANG(CANG(CENTER))
37          RN=SQRT(1-2*AK*TMP1*GI+(TMP1*GI)**2)
38          RADIUS=RN/(1+GI*TMP2)
39          PRINT *
40          PRINT *,G,' DB GAIN CIRCLE .   RADIUS= ',RADIUS
41          PRINT *,'    CENTER.MAG = ',DISTAN,'   CENTER.ANG = ',ANGLE
42          G=G+STEP
43          IF(G.LE.STOP) GO TO 43
44          GO TO 3
45    103   RETURN
46          END


1           SUBROUTINE NOISEF
2           COMPLEX GO
3           PRINT *,' ENTER FMIN(DB), GO.MAG, GO.ANG, RN'
4           READ *, FMIN,GOMAG,GOANG,RN
5           FMIN=10**(FMIN/10)
6           R=RN/50
7     4     PRINT *
8           PRINT *,'ENTER NOISE FIG. IN DBS -- START,STOP,STEP'
9           READ *,START,STOP,STEP
10          F=START
11    14    FI=10**(F/10)
12          GO=CMPLX(GOMAG*COS(RAD(GOANG)),GOMAG*SIN(RAD(GOANG)))
13          TMP=CABS(1+GO)**2
14          ANI=(FI-FMIN)*TMP/(4*R)
15          CFIMAG=GOMAG/(1+ANI)
16          CFIANG=GOANG
17          RFI=SQRT((ANI)**2+ANI*(1-GOMAG**2))/(1+ANI)
18          PRINT *, F ,' DB NOISE FIGURE CIRCLE'
19          PRINT *
20          PRINT *,'     CENTER.MAG= ',CFIMAG,'  CENTER.ANG = ',CFIANG
21          PRINT *,'          RADIUS = ',RFI
22          PRINT *
23          F=F+STEP
24          IF (F.LE.STOP) GO TO 14
25          PRINT *,' ENTER 1 TO CONTINUE, 0 TO EXIT'
26          READ *,J
27          IF(J.EQ.1) GO TO 4
28          RETURN
29          END
```

## Appendix B  UM-MAAD

```
1          SUBROUTINE REFLEC (S11,S21,S12,S22)
2          COMPLEX S11,S21,S12,S22,SN,SD,GAMMA,G
3          PRINT *
4     5    PRINT *,' ENTER    1 FOR INPUT REFLECTION COEFFICIENT '
5          PRINT *,'          2 FOR OUTPUT REFLECTION COEFFICIENT '
6          PRINT *,'          3 TO EXIT '
7          READ *,L
8          IF (L.EQ.3) GO TO 55
9          PRINT *,'   ENTER GAMMA TERMINATION (MAG,ANG)'
10         READ *,GAMMAG,GAMANG
11         GAMANG=RAD(GAMANG)
12         GAMMA=CMPLX(GAMMAG*COS(GAMANG),GAMMAG*SIN(GAMANG))
13         GO TO (25,35) L
14         PRINT *,' TRY AGAIN'
15         GO TO 5
16    25   SN=S11
17         SD=S22
18         PRINT *,'   ****** THE INPUT REFLECTION COEFFICIENT ******'
19         GO TO 45
20    35   SN=S22
21         SD=S11
22         PRINT *,'   ****** THE OUTPUT REFLECTION COEFFICIENT ******'
23    45   G=SN+S12*S21*GAMMA/(1-SD*GAMMA)
24         GMAG=CABS(G)
25         GANG=ANG(CANG(G))
26         PRINT *,'   GAMMA.MAG = ',GMAG,'GAMMA.ANG = ',GANG
27         PRINT *
28         GO TO 5
29    55   RETURN
30         END


1          SUBROUTINE CONVER
2          COMPLEX Z,ZZ,YY,GAMMA,GAM
3     6    PRINT *
4          PRINT *,'   ENTER 1 FOR Z TO GAMMA'
5          PRINT *,'         2 FOR GAMMA TO Z AND Y'
6          PRINT *,'         3 TO EXIT '
7          READ *,I
8          IF (I.EQ.3) GO TO 56
9          GO TO (26,36) I
10         PRINT *,' TRY AGAIN'
11         GO TO  6
12    26   PRINT *,'ENTER NORMALIZED Z (REAL,IMAG)'
13         READ *,ZRE,ZIM
14         Z=CMPLX(ZRE,ZIM)
15         GAMMA=(Z-1)/(Z+1)
16         GMAG=CABS(GAMMA)
17         GANG=ANG(CANG(GAMMA))
18         PRINT *,'   GAMMA.MAG = ',GMAG,'   GAMMA.ANG  = ',GANG
19         PRINT *
20         GO TO 46
21    36   PRINT *,' ENTER GAMMA (MAG,ANG)'
22         READ *,GAMAG,GAANG
23         GAM=CMPLX(GAMAG*COS(RAD(GAANG)),GAMAG*SIN(RAD(GAANG)))
24         ZZ=(1+GAM)/(1-GAM)
25         YY=1/ZZ
26         PRINT *,'   IMPEDANCE Z = ',ZZ
27         PRINT *,'   ADMITTANCE Y =',YY
28         PRINT *
29    46   GO TO 6
30    56   RETURN
31         END
```

238                                              UM-MAAD     Appendix B

```
1       FUNCTION CANG(W)
2       COMPLEX W
3       X=REAL(W)
4       Y=AIMAG(W)
5       CANG=ATAN2(Y,X)
6       RETURN
7       END
```

```
1       FUNCTION ANG(RAD)
2       PI=3.14159265
3       ANG=RAD*180/PI
4       RETURN
5       END
```

```
1       FUNCTION RAD(ANG)
2       PI=3.14159265
3       RAD=ANG*PI/180
4       RETURN
5       END
```

A typical run of UM-MAAD is illustrated below. UM-MAAD is used to calculate the results given in Examples 3.7.1 and 3.8.1. The menu from subroutine SELECT is printed only once and is shown as MENU SELECTION in subsequent appearances. The user's entries are shown after the start of entry sign, >.

## TYPICAL RUN

```
ENTER MAG. AND ANG. OF S11,S21,S12,S22
> .641  -171.3  2.058  28.5  .057  16.3  .572  -95.7


    S11=               S21=              S12=            S22=
    .641  -171.300    2.058   28.500    .057   16.300   .572   -95.700

    ENTER 1 FOR STABILITY CIRCLES,K, AND DELTA
          2 FOR UNILATERAL FIGURE OF MERIT
          3 FOR UNILATERAL POWER GAIN CIRCLES
          4 FOR SIMULTANEOUS CONJUGATE MATCH
          5 FOR POWER & AVAIL. POWER GAIN CIRCLES
          6 FOR NOISE FIGURE CIRCLES
          7 FOR INPUT/OUTPUT REFLECTION COEFFICIENTS
          8 FOR CONVERSION BETWEEN GAMMA AND Z
          9 TO EXIT


   > 1
```

Appendix B    UM-MAAD                                                          **239**

```
           DELTA.MAG=    .301    DELTA.ANG= 109.865         K=    1.504

   UNCONDITIONALLY STABLE TRANSISTOR
   ---------------------------------
   **** INPUT STABILITY CIRCLE ****

   RADIUS=  .36655562    CENTER.MAG= 1.4955822    CENTER.ANG=  177.29895
   OUTSIDE IS STABLE

   **** OUTPUT STABILITY CIRCLE ****

   RADIUS=  .49637499    CENTER.MAG= 1.6550584    CENTER.ANG=  103.93965
   OUTSIDE IS STABLE


MENU SELECTION
> 2

     ****** UNILATERAL FIGURE OF MERIT =    .10851129

    -.89480235     DB < GT/GTU <   .99768301      DB



MENU SELECTION
> 4

   **** COMPLEX CONJUGATE MATCH AT THE INPUT PORT
        GAMMAS.MAG =   .76194154       GAMMAS.ANG =    177.29895

   **** COMPLEX CONJUGATE MATCH AT THE OUTPUT PORT
        GAMMAL.MAG =   .71838430       GAMMAL.ANG =    103.93965



MENU SELECTION
> 8


   ENTER 1 FOR Z TO GAMMA
         2 FOR GAMMA TO Z AND Y
         3 TO EXIT



> 2

ENTER GAMMA (MAG,ANG)

>  .762   177.3


   IMPEDANCE  Z = (  .13514743    , .23136132-001)
   ADMITTANCE Y = ( 7.1886514    ,-1.2306381    )


   ENTER 1 FOR Z TO GAMMA
         2 FOR GAMMA TO Z AND Y
         3 TO EXIT


> 3

MENU SELECTION
> 5

   UNCOND. STABLE TRAN. GPMAX= 11.381487     DB

      ENTER 1 FOR OPERATING GAIN CIRCLES.
            2 FOR AVAILABLE GAIN CIRCLES.
            3 FOR EXIT
```

```
> 1

ENTER GAIN IN DBS -- START,STOP,STEP

> 9  10  2

       ****** OPERATING POWER GAIN CIRCLES ******

    9.0000000     DB GAIN CIRCLE .   RADIUS=     .43090543
       CENTER.MAG =    .50827681        CENTER.ANG =    103.93965

          ENTER 1 FOR OPERATING GAIN CIRCLES.
                2 FOR AVAILABLE GAIN CIRCLES.
                3 FOR EXIT


> 3

MENU SELECTION

> 7

          ENTER   1 FOR INPUT REFLECTION COEFFICIENT
                  2 FOR OUTPUT REFLECTION COEFFICIENT
                  3 TO EXIT


> 1

ENTER GAMMA TERMINATION (MAG,ANG)

> .36   47.5

       ****** THE INPUT REFLECTION COEFFICIENT ******
       GAMMA.MAG =     .62902135    GAMMA.ANG =   -175.51289



@FIN
```

# INDEX

**ABCD parameters, 2–4, 24–25**
  matrix, 1–3
Active bias:
  BJT, 130
  GaAs FET, 133
Added power, 207
Admittance coordinates Smith chart (*see* Smith chart)
Admittance parameters, 2–4, 24–26
  matrix, 2
AM to PM conversion, 181
Amplifier:
  balanced, 159–60
  bandwidth, 170–74
  broadband, 154–69
  feedback, 102, 155, 160–66
  high-power, 174–87
  low-noise, 140–54
  stability (*see* Stability)
  tuning, 169–70
  two-stage, 187–88
AMPSYN, 156, 193
Attenuation constant, 5
Available noise power, 141
Available power, 15–18, 21, 85–86, 92, 206–7
Available power gain (*see also* Power gain):
  circles (*see* Constant-gain circles)
  maximum, 32
Average power, 14

**Balanced amplifier, 159–60**
Balanced shunt stubs, 75–76, 78–79

Bandwidth:
  analysis, 170–74
  inherent, 171–72
  reduction factor, 174
Beta cutoff frequency, 28–29, 33
Bias circuit (*see* dc bias)
BJT:
  bias (*see* dc bias)
  characteristics, 31–34, 139
  figure of merit, 32
  junction temperature, 181
  model, 31–32
Boltzman constant, 140
Broadband design, 154–69

**CAD, 155, 166, 217–40**
Capacitor:
  bypass, 130
  chip, 75
  coupling, 75
Cascade, 3–4, 11
Chain parameters, 2
  matrix, 2–3
Chain scattering parameters, 11
  matrix, 8, 11–12
Characteristic impedance:
  complex, 5
  microstrip, 68–69
  real, 13
  transmission line, 5, 7, 13
Chip, 23, 27–28

**242**  Index

Class:
  A operation, 130, 133, 139
  AB operation, 130, 133
  B operation, 130, 133
COMPACT, 217–22
Compensated matching networks, 155, 158
Complex propagation constant, 5
Compressed Smith chart (*see* Smith chart)
Compression point (*see* Gain compression point)
Computer-aided design (*see* CAD)
Conduction loss, 72
Conjugate match:
  maximum gain, 112–14
  simultaneous (*see* Simultaneous conjugate match)
Constant-conductance circles, 46
Constant-gain circles:
  available power gain, 123
    potentially unstable, 123–25
    unconditionally stable, 123
  operating power gain, 119–25
    potentially unstable, 123–25
    unconditionally stable, 119–23
  transducer power gain, 102–10, 114–19
    bilateral case, 114–19
      potentially unstable, 118–19
      unconditionally stable, 114–15
    unilateral case, 102–10
      potentially unstable, 106–10
      unconditionally stable, 103–6
Constant-reactance circles, 44, 46
Constant-resistance circles, 44, 46
Constant-susceptance circles, 46
Contact potential, 213
  resistance, 27
Conversions, 23–26
Coupler, 159

**dc bias:**
  BJT, 125–30
  GaAs FET, 131–33
  networks, 128–33
  operating point, 130–31, 133
  stability, 126–27
Device-line characterization, 205
Dielectric constant:
  effective, 68
  relative, 67
Dielectric loss, 72
Dielectric substrate, 67
Dispersion, 71–72
Distortion, 177, 180
Dynamic range, 175, 180
  free spurious, 179–80

**Electrical length, 5, 13**
Electron transit time, 35
  saturation drift velocity, 35

Ell matching sections, 55–56
Error function, 219–20, 225

**Fano, 167**
Feedback:
  negative (*see* Negative feedback)
  series, 161
  shunt, 161
Figure of merit, 32, 114
Flow graph (*see* Signal flow graphs)

**GaAs FET:**
  bias (*see* dc bias)
  characteristics, 34–36, 139
  junction temperature, 181
  model, 34–36
Gain-bandwidth, 32
Gain circles (*see* Constant-gain circles)
Gain compression point, 175
Gate length, 35
Gyromagnetic ratio, 213

***h* parameters (*see* Hybrid parameters)**
$h_{FE}$, 32, 126–27
Hybrid combiner/divider, 181–82
Hybrid parameters, 2, 4, 24–25
  matrix, 2
Hybrid-$\pi$ model, 31

$I_{CBO}$, **126–27**
IGFET, 34
Impedance coordinates Smith chart (*see* Smith chart)
Impedance matching networks (*see* Matching)
Impedance parameters, 1–4, 24–25
  matrix, 2
Incident:
  power, 14
  voltage, 14
  wave, 5, 9–10, 14, 20
Incoming wave, 5
Indefinite scattering matrix, 29
Input reflection coefficient (*see* Reflection coefficient)
Input stability circle, 96
Insulated-gate field-effect transistor (*see* IGFET)
Interdigitated, 159
Intermodulation distortion:
  products, 178–79
  third-order, 178
Interstage design, 155–56, 187

**JFET, 34**
Junction field-effect transistor (*see* JFET)

**Lange coupler, 159–60**
Large signal:
  characterization, 174–75

Index

measurements, 176–78, 203–8
scattering parameters, 174–78
Low-noise amplifier (*see* Amplifier)

**Mason's rule, 83–86**
Matched line, 8
  impedance, 16
  load, 85–86
  termination, 10, 21
Matching:
  network, 55–56
  network design, 55–67, 74–80, 116–19,
    147–49, 157–59, 185–86, 203
Maximum available gain (*see* Power gain)
Maximum available noise power, 140
Maximum frequency of oscillation ($f_{max}$), 32–35
Maximum power gain (*see* Power gain)
Maximum stable gain, 114
MESFET, 34
Metal oxide semiconductor field-effect
    transistor (*see* MOSFET)
Metal semiconductor field-effect transistor (*see*
    MESFET)
Microstrip:
  attenuation, 72–73
  capacitance, 68
  characteristic impedance, 68–69
  definition, 67
  dispersion, 71–72
  effective dielectric, 68
  effective width, 71
  field configuration, 67
  geometry, 67
  losses, 72–73
  matching network design, 74–80, 116–19 (*see
    also* Matching)
  phase velocity, 68
  quality factor, 73–74
  radiation factor, 74
  wavelength, 68–70
Microwave amplifier:
  block diagram, 55
  schematic, 78–80
  signal flow graph, 83
Microwave transistor:
  BJT, 31–33
  characteristics, 31–36
  GaAs FET, 34–36
Minimum noise figure (*see* Noise)
MOSFET, 34

**Negative feedback, 102, 155, 160–66**
Negative resistance, 48–49
Negative resistance oscillators (*see* Oscillators)
Nepers, 5
Noise:
  bandwidth, 140
  extrinsic, 36
  figure, 141–44, 151–54

  figure circles, 142–47
  figure parameters, 143
  intervalley, 36
  intrinsic, 36
  Johnson, 140
  measurement, 143
  minimum, 141–54
  parameters, 143, 145, 148, 152
  power, 140, 175
  resistance, 142
  resistor, 140
  shot, 34
  temperature, 140
  thermal, 140, 175
  two-stage amplifier, 141–42
  voltage, 140
  white, 140
Normalizing impedance, 15–22
  admittance, 46
  resistance, 15, 22
$n$-port network, 19–22
$n$-way amplifier, 182–84
$n$-way hybrid combiner/divider, 182–84

**Open-circuited line, 8**
  shunt stub, 75, 78, 116–19, 148, 185
Operating point (*see* dc bias)
Operating power gain (*see* Power gain)
  circles (*see* Constant-gain circles)
Optimization, 217–28
Optimum:
  noise figure, 142
  terminations, 103, 105
Oscillation conditions, 195–96
Oscillators:
  BJT, 210, 213
  Clapps, 209
  Colpitts, 209
  configurations, 208–14
  design procedure, 200
  GaAs FET, 210–13
  Hartley, 209
  large-signal measurements, 203–9
  maximum power, 197–98
  negative resistance, 194–208, 210
  one-port, 194–99
  reverse channel, 200, 210–12
  terminating network, 199–203
  two-port, 199–203
  varactor tuned, 213–14
  YIG tuned, 212–13
Outgoing wave, 5
Output reflection coefficient (*see* Reflection
  coefficient)
Output stability circle, 96, 124–25

**Packaged, 23, 27–28**
Packaged capacitance, 27
  inductance, 27

Paralleling, 180–81
Parameters conversions, 23–26
Parasitics, 32
Phase velocity, 5, 7
Phasor, 4
Potentially unstable:
  bilateral, 118–19, 123–25
  unilateral, 106–10
Power amplifier, 174–86
Power combiners and dividers:
  hybrid, 181–82
  $n$-way, 182–84
  Wilkinson, 181–83
Power delivered to the load, 16–17, 85–86, 92
Power gain:
  available, 92, 123
  maximum available, 32, 121, 141
  maximum operating, 120–21
  maximum stable, 114
  maximum transducer, 114, 121
  maximum unilateral transducer, 94, 106
  operating, 92, 119–21
  transducer, 17, 22, 32, 86, 114
  unilateral transducer, 93–94, 102
Propagation constant, 5

**Q, 168–69, 171**
Quarter-wave transformer, 8, 78, 147, 185
Quasi-TEM, 67–68, 71
Quiescent point, 130–33

**Radiation factor, 74**
Radio frequency choke (RFC), 129–30
Reference:
  impedance (see Normalizing impedance)
  plane, 12–13
  resistance (see Normalizing resistance)
Reflected:
  power, 16–17, 21
  voltage, 14
  wave, 5–6, 9–10, 14, 20
Reflection coefficient:
  definition, 5, 9
  input, 10, 22, 84–85
  load, 5, 9
  noise, 142–44, 188
  output, 10, 85
  plane, 43
  power, 176–77, 187
  simultaneous conjugate match, 112–14
  transmission line, 5–6, 8
  two-stage design, 187–88
Resonance line width, 213
Reverse channel, 200, 210–12

**Scattering matrix, 8, 10, 23**
Scattering parameters:
  conversions, 24–25
  definitions, 10, 16–18
  generalized, 19–22
  indefinite, 29
  large-signal, 174–78
  matrix, 8, 10, 23
  measurement, 10–11
  $n$-port, 20
  properties, 13–18
  transistors, 23, 26–31
Scattering transfer parameters (see Chain scattering parameters)
Schottky, 34, 213
Shifting reference planes, 12–13
Short-circuit current gain (see $h_{FE}$)
Short-circuited line, 8
  shunt stub, 75–78, 147–48
Shot noise, 34
Signal flow graphs:
  applications, 84–87
  branch, 80–81
  generator, 81–82
  input reflection coefficient, 84–85
  load, 82–83
  loops, 84
  microwave amplifier, 83
  output reflection coefficient, 85
  path, 83
  theory, 80–87
  two-port, 81
Simultaneous conjugate match, 112–14
Smith chart:
  admittance or $Y$ chart, 45–47
  compressed, 48
  design, 55–66
  gain circles (see Constant-gain circles)
  impedance calculation, 49–51
  impedance or $Z$ chart, 45–47
  negative resistance, 48–49
  network characteristics, 53–54
  normalized impedance and admittance or $ZY$ chart, 51–54, 55–66
  theory, 43–51
Stability:
  analysis, 95–100, 164–65
  circles, 96–98, 101, 124–25, 201
  conditions, 98–100
  potentially unstable, 95–96, 99–100
  unconditionally stable, 95–100
Stability factors, 127
Standard temperature, 140
Standing wave ratio (see VSWR)
Strip conductor, 67
Stub (see Short- or open-circuited line)
Substrate, 67
SUPER-COMPACT, 222–28

**TEM mode, 67**
Terminating matching network, 199–203
Thermal noise, 34
Thévenin, 16–17, 20

# Index

Third-order intercept point, 178–79
Three-port, 29, 31
Transconductance, 163
Transducer cutoff frequency, 28
Transducer power gain, 17, 22, 32, 86 (see also Constant-gain circles)
Transfer or $T$ parameters (see Chain scattering parameters)
Transmission coefficient:
  forward, 10, 17, 28
  reverse, 10, 28
Transmission line, 4–8
  input impedance, 6–7, 49–50
  lossless, 7, 13–14
  matched, 8
  open-circuited, 8
  quarter-wave, 8
  short-circuited, 8
  uniform, 5, 36
Traveling waves, 4–7, 9, 13
Tuning factor, 170
Two-port network:
  cascade, 11
  noise, 140–42
  parameters conversions, 22–25
  representations, 1–4, 9, 13
Two-stage amplifier:
  design, 187–88
  high-gain, 187
  high-power, 187
  low-noise, 188

**UM-MAAD, 230–40**
Unconditionally stable:
  bilateral, 114–23
  conditions, 98–100
  unilateral, 103–6
Unilateral figure of merit, 111–12
Unilateral transducer power gain (see Power gain)
Unit diagonal matrix, 23
Unstable two-port (see Stability)

**Varactor diode, 213–14**
Voltage gain, 87
Voltage standing-wave ratio (see VSWR)
VSWR, 7–8, 50–51

**Wavelength:**
  free space, 68
  microstrip, 68
Wilkinson's coupler, 181–83

$y$ parameters (see **Admittance parameters**)
YIG sphere, 212–13

$z$ parameters (see **Impedance parameters**)

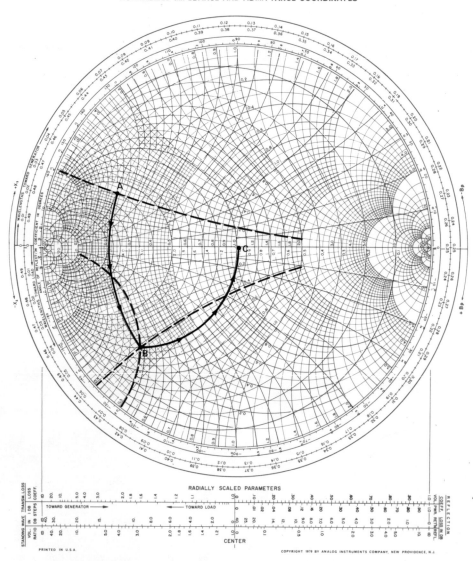

**Figure 2.4.9b**